Water, Climate Change and the Boomerang Effect

In line with COP21 agreements, state-led climate change mitigation and adaptation actions are being undertaken to transition to carbon-neutral, green economies. However, the capacity of many countries for action is limited and may result in a 'boomerang effect', defined as the unintended negative consequences of such policies and programmes on local communities and their negative feedbacks on the state. To avoid this effect, there is a need to understand the policy drivers, decision-making processes, and impacts of such action, in order to determine the ways and means of minimizing negative effects and maximizing mutually beneficial policy outcomes.

This book directly engages the policy debates surrounding water resources and climate actions through both theoretical and comparative case studies. It develops the 'boomerang effect' concept and sets it in relation to other conceptual tools for understanding the mixed outcomes of state-led climate change action, for example, 'backdraft' effect and 'maldevelopment'. It also presents case studies illustrative of the consequences of ill-considered state-led policy in the water sector from around the world. These include Africa, China, South Asia, South America, the Middle East, Turkey and Vietnam, and examples of groundwater, hydropower development and forest hydrology, where there are often transboundary consequences of a state's policies and actions. In this way, the book adds empirical and theoretical insights to a still developing debate regarding the appropriate ways and means of combating climate change without undermining state and social development.

Larry A. Swatuk is a Professor in the School of Environment Enterprise and Development, University of Waterloo, Canada and Extraordinary Professor in the Institute for Water Studies, University of the Western Cape, South Africa.

Lars Wirkus is a Researcher and Head of Section, Data and GIS at the Bonn International Center for Conversion (BICC), Germany.

Earthscan Studies in Water Resource Management

For more information and to view forthcoming titles in this series, please visit the Routledge website: www.routledge.com/books/series/ECWRM/

Water, Climate Change and the Boomerang Effect

Unintentional Consequences for Resource Insecurity

Edited by Larry A. Swatuk and Lars Wirkus

LONDON AND NEW YORK

from Routledge

First published 2018
by Routledge
2 Park Square, Milton Park, Abingdon, Oxon OX14 4RN

and by Routledge
52 Vanderbilt Avenue, New York, NY 10017

First issued in paperback 2020

Routledge is an imprint of the Taylor and Francis Group, an informa business

British Library Cataloguing-in-Publication Data
A catalogue record for this book is available from the British Library

Library of Congress Cataloging-in-Publication Data
A catalog record has been requested for this book

ISBN 13: 978-0-367-58814-4 (pbk)
ISBN 13: 978-1-138-55609-6 (hbk)

Typeset in Goudy
by Wearset Ltd, Boldon, Tyne and Wear

Contents

Illustrations

Figures

Tables

About the authors

Alhassan Lansah Abdulai pursued a Master of Science (MSc) programme in Agricultural Science and Resource Management in the Tropics and Sub-Tropics at the University of Bonn in Germany. He also obtained a PhD in Climate Change Adaptation from the University of Hohenheim in Stuttgart, Germany. Currently he works with the Council for Scientific and Industrial Research (CSIR), Ghana and focuses on Climate Change Adaptation at the Savanna Agricultural Research Institute (SARI) in Tamale Ghana.

Romy Buchner is a graduate student in the Masters of Development Practice programme at the University of Waterloo in Canada. Her current research focuses on Maternal Health and Gender Equality.

Chieh Cheng holds a Master of Development Practice degree from the University of Waterloo, Canada.

Mary Crawford holds a Master of Development Practice degree from the University of Waterloo, Canada.

Geoffrey D. Dabelko is Professor and Associate Dean at Ohio University's Voinovich School of Leadership and Public Affairs and is Advisor to the Wilson Center's Environmental Change and Security Program.

Frances Delaney holds a Master of Climate Change degree from the University of Waterloo, Canada.

Jason Durst holds a Master of Development Practice degree from the University of Waterloo, Canada.

Zoya Khan holds a Master of Development Practice degree from the University of Waterloo, Canada.

Ricarda Ines Konwiarz is a student in the Master's Programme in Politics and Public Administration at the University of Konstanz in Germany. Her research focuses on topics linked to global governance, in particular global health issues and access to essential medicines.

Florian Krampe (PhD, Uppsala) is a political scientist based at the Stockholm International Peace Research Institute (SIPRI) where he works on climate security and sustaining peace. Krampe specializes in international relations, peace and conflict research and international security. His primary academic interest is the foundations of peace and security, especially the processes of building peace after armed conflict. Currently, he focuses on climate security and the post-conflict management of natural resources with a specific interest in the ecological foundations for a socially, economically and politically resilient peace.

Sonya Deborah Krause holds a Master of Climate Change degree from the University of Waterloo, Canada.

Kaylia Little graduated with a Master of Development Practice from the University of Waterloo and a BAH in Global Development Studies from Queen's University in Canada. Currently, she works with the Ontario Native Women's Association in Thunder Bay. She will begin her PhD in Sustainability Management at the University of Waterloo in Fall 2018.

Stephen Little graduated with a Masters of Development Practice from the University of Waterloo in Canada, with an interest in Economic Development and Entrepreneurship. Currently he is the Product Manager at Trylon (Elmira, ON).

Yuye Li holds a Master of Development Practice degree from the University of Waterloo, Canada.

Laura Maxwell holds a Master of Development Practice degree from the University of Waterloo, Canada.

Mridula Nair holds a Master of Development Practice degree from the University of Waterloo, Canada.

Vidya Nair holds a Master of Development Practice degree from the University of Waterloo, Canada. With substantial direct UN, civil society and field experience in Southern Africa and Southeast Asia, her research and policy specialisms include, water sanitation and hygiene (WASH), capacity building and women's empowerment. More specifically, her current work focuses on reviewing policy recommendations in Ontario's education sector.

Liam Neumann, holds a Master of Development Practice degree from the University of Waterloo, Canada.

Meaghan Parker is Senior Writer/Editor at the Wilson Center's Environmental Change and Security Program.

Barbara Pinto holds a Masters of Climate Change degree from the University of Waterloo and a Bachelor's of Environmental Science degree from Queen's University. Currently, she works with the Lake Simcoe Region Conservation Authority, where she is developing a more refined understanding of the impacts of climate change on the Lake Simcoe watershed's biodiversity, water quality and water quantity.

Rija Rasul graduated with a Masters of Development Practice from the University of Waterloo in Canada. She currently works with the City of Vaughan's Environmental Sustainability Department, and her research interests lie in the synergies between climate change adaptation and mitigation.

Kadra Rayale graduated with a Master's in Development Practice from the University of Waterloo. Currently based in Ottawa, her research focuses on forced and voluntary migration trends, climate change and innovations in international development.

Nikita Yasmin Shah holds a Master of Global Governance degree from the Balsillie School of International Affairs in Waterloo, Canada.

Luis Paulo Batista da Silva is a PhD in Geography at Federal University of Rio de Janeiro, Brazil. Currently, his research investigates the role of sub-catchments at La Plata's transboundary river basins governance. He is also a researcher at Grupo Retis, a research team interested at South American border regions (www.retis.igeo.ufrj.br/)

Ana Smith holds a Master of Development Practice degree from the University of Waterloo, Canada.

Sebastiaan Soeters is an independent research consultant working on issues relating to various forms of climate change policy and practice, and how they relate to new dynamics of conflict and exclusion in African drylands. He has a special interest in farmer–pastoral relations.

Stephanie Solomon holds a Master of Development Practice degree from the University of Waterloo, Canada.

Bejoy K. Thomas has a background in Economics and Development Studies. He is currently Fellow with the Centre for Environment and Development, Ashoka Trust for Research in Ecology and the Environment (ATREE), Bangalore, India. His research is problem driven and interdisciplinary, oftentimes in close collaboration with environmental scientists and engineers, focusing on land and water resources in rural and peri-urban areas. He has been a Visiting Fellow at the Water Institute, University of Waterloo, Canada (2016) and the Liu Institute for Global Issues, The University of British Columbia, Canada (2017).

Bojian Zhang graduated with a Master of Development Practice degree from the University of Waterloo, Canada. Currently, he works with Geely, a Chinese automobile company. His research focuses on strategies for eliminating air pollution and poverty in China.

Zhe, Zhang holds a Master of Development Practice degree from the University of Waterloo, Canada.

Preface

This project began its life as a cross-Atlantic conversation regarding the emergent global discussion about the potential for climate change actions to result in unexpected negative effects. Numerous projects and perspectives were at play around similar phenomena: new forest management strategies in support of global goals regarding carbon emissions; crop switching for global food and energy markets; national strategies for water, energy and food security. Climate actions were interpreted differently: one person's 'climate security', or 'energy security', was another person's 'land grab' or 'water grab'. After watching a webinar on 'backdraft', which included a discussion of geo-engineering, one of us remarked to the other 'What isn't "backdraft"?' With financial assistance from the University of Waterloo's EU-IRPG fund and a generous grant from the University of Waterloo's Water Institute, we constituted several workshops devoted to our idea of the 'boomerang effect'. To us, 'backdraft' is an inaccurate image of the actual causal chain. More accurate, in our view, is the boomerang. An aboriginal hunting weapon, a boomerang is deliberately thrown by a hunter at his/her intended prey. Missing the target, results in the boomerang returning to the hand of the hunter. Policymakers mimic hunters in crafting actions designed to have a particular impact. Unlike a hunter who has a single target, policy in the social world rarely has a single target or single impact. Often times, the impact is not in any way related to that intended. In relation to climate action (and much development policy generally), policymakers in our view fail to consider all factors affecting the likelihood of success. So, global policies designed to sequester carbon in support of 'planetary boundaries' can seriously negatively affect local people dependent upon forests. This effect can boomerang back to the state actor who, if not expecting such an impact, will also be negatively affected. In effect, if you are not expecting the boomerang to return, you can be struck with it. A good hunter should consider all eventualities. By separating local-level side effects of climate action from boomerang effects we hope to help policymakers to make better policy. By 'better' we mean not only policy that hits its intended target, but which maximizes social, economic, political and environmental benefits.

In support of our goal, we developed the concept idea during Swatuk's 2015 sabbatical at the Bonn International Center for Conversion (BICC). The aforementioned financial support enabled us to hold an initial meeting in September 2016 at the University of Waterloo. In preparation for this meeting, Swatuk tasked his graduate class in Water and Security with writing background papers in line with the 'boomerang effect' concept. These papers were presented at the Waterloo meeting and later also at the annual meeting of the International Conference on Sustainable Development (ICSD) at Columbia University in New York. They have all been substantially revised and brought more into line with the concept as it has evolved over the subsequent 18 months. The initial meeting in Waterloo involved the authors of Chapter 1 in this collection as well as a number of other colleagues. In September 2017, Swatuk and Wirkus, along with Bejoy K. Thomas, Florian Krampe and Luis Paulo Batista da Silva, attended the Water Security conference in Cologne, Germany. Here we presented the refined theoretical framework and a case study from South Africa. Following this meeting, we held a small workshop at BICC to further work on the project and chart a way forward. This book is a first attempt to engage a wider audience in what we believe to be a very important discussion regarding the ways and means of devising appropriate climate action. This book marks the starting point for further research. In particular, we are interested not only in refining the theoretical framework, but in determining and devising appropriate and applicable tools to enable decision makers to make better decisions – highly accurate geointelligence solutions, big data and earth observation analytics service provisioning platforms, scenario building mechanisms, and so on.

The editors would like to thank a number of people and organizations for their support of the project of which this book is the first product. First and foremost, we thank our home organizations: the School of Environment, Enterprise and Development (SEED) at the University of Waterloo and the Bonn International Center for Conversion (BICC). We especially thank the Water Institute and the University of Waterloo EU-IRPG fund for making the project possible. We also thank those who attended the initial workshop in Waterloo, in particular Alain Nimubona and Natasha Tang Kai. Valuable insights were also gained from meetings at UNU-EHS in Bonn. We are grateful for the support, patience and guidance of Amy-Louise Johnston and Tim Hardwick. Lastly we thank our spouses, Corrine and Thekla, for their unflagging support.

Larry Swatuk, Waterloo, Canada
Lars Wirkus, Bonn, Germany
10 April 2018

Acronyms

ACCES	Africa Climate Change Environment and Security
AEAs	Agricultural Extension Agents
AF	Agroforestry
AFI	Artificial Floating Island
AGRA	Alliance for Green Revolution in Africa
BBC	British Broadcasting Corporation
BMGF	Bill and Melinda Gates Foundation
BMI	Business Monitor International
BNDES	Brazilian Development Bank
BRICS	Brazil, Russia, India, China and South Africa
CA	Conservation Agriculture
CBA	Cost Benefit Analysis
CCMCC	Climate Change Mitigation Conflict and Cooperation
CDM	Clean Development Mechanism
CHWs	Community Health Workers
CO_2	Carbon Dioxide
COP21	Twenty-first annual Conference of the Parties
CORAF/WECARD	West and Central African Council for Agricultral Research and Development
CPC	Communist Party of China
CPC	People's Committee
CPPCC	Chinese People's Political Consultative Congress
CSIR-SARI	Council for Scientific and Industrial Research through the Savanna Agricultural Research Institute
DALYs	Disability Adjusted Life Years
DFID	Department for International Development of the United Kingdom
DOA	Department of Agriculture
DSI	State Water Utility
EAPP	East African Power Pool
ECA	Export Credit Agencies
EIA	Environmental Impact Assessments
EJOLT	Environmental Justice Organization, Liabilities and Trade

ENRACCA-WA	Enhancing Resilience and Adaptive Capacity to Climate Change through Integrated Land, Water and Nutrient Management in Semi-Arid West Africa
EPRDF	Ethiopian People's Revolutionary Democratic Front
EPSEMP	Ethiopian Power System Expansion Master Plan
ESIA	Environment and Social Impact Assessment
FAO	Food and Agriculture Organization
FGD	Focus Group Discussions
FPIC	Free, prior and informed consent
G7	Group of Seven (Highly Industrialized Countries)
GAA	Guarani Aquifer Agreement
GAP	Güneydoğu Anadolu Projesi (South-Eastern Anatolia Project)
GAP RDA	GAP Regional Development Administration
GAS	Guarani Aquifer System
GBM	Ganges–Brahmaputra–Meghna
GCF	Green Climate Fund
GDP	Gross Domestic Product
GERD	Grand Ethiopian Renaissance Dam
GHG	Green House Gas
GMB	Ganges–Meghna–Bramaputra
GoG	Government of Ghana
GOLD	General Organization for Land Development
GT CO_2	Gigatons of Carbon dioxide
GWh	Gigawatt hours
IAEA	Germany's Federal Institute for Geosciences and Natural Resources
IBA	Important Bird Area
IBAMA	Brazilian Institute of Environment and Renewable Natural Resources
ICBC	Industrial and Commercial Bank of China
ICOLD	International Commission on Large Dams
ICSSI	Iraqi Civil Society Solidarity Initiative
IDPs	Internally Displaced People
IEA	International Energy Agency
IFAD	International Fund for Agricultural Development
IIRR	International Institute of Rural Reconstruction
ILO	International Labour Organization
IMEP	Israel Ministry of Environmental Protection
INDC	Intended Nationally Determined Contribution
INM	Integrated Nutrient Management
INPA	National Institute for Research in the Amazon
INRM	Integrated Natural Resource Management
IPCC	Intergovernmental Panel on Climate Change
IPM	Integrated Pest Management

IPT	Intermittent Preventive Treatment
ISFM	Integrated Soil Fertility Management
ISIS	Islamic State of Iraq and Syria
ITNs	Insecticide Treated Nets
IWMI	International Water Management Institute
IWRM	Integrated Water Resource Management
I(G)WRM	Integrated Water Resource Management
JRB	Jordan River Basin
JWC	Joint Water Commission
KHRP	Kurdish Human Rights Project
kWh	Kilowatt Hour
LLSE	Local-level Side Effects
LURC	Land Use Rights Certificates
MAB	Movement of People Affected by Dams
MDGs	Millennium Development Goals
MDTX	Movement for the Development of the Trans-Amazon and Xingu
MESTI	Ministry of Environment, Science, Technology and Innovation
MMA	Minister of Environment
MOEF	Ministry of Environment and Forests
MoFA	Ministry of Food and Agriculture
MoFA	Ministry of Foreign Affairs
MoU	Memorandum of Understanding
MPEG	Emílio Goeldi Museum of Para'
MWh	Megawatt Hours
NAEA	Nucleus for High-Level Studies of Amazonia
NAES	Nyankpala Agricultural Experimental Station
NATO	North Atlantic Treaty Organization
NDC	Nationally Determined Contributions
NE	Norte Energia
NGO	Non-governmental Organization
NPC	National People's Congress
NRAP	National REDD+ Action Plan
NRLP	National River Linking Project
NRN	National REDD Network
NWM	National Water Mission
NWO	Dutch Organization for Scientific Research
NWRC	National Water Resources Council
OA	Organic Agriculture
OAS	Organization of American States
OECD	Organization for Economic Co-operation and Development
PA	Palestinian Authority
PA	Precision Agriculture

PES	Payment for Ecosystem Services
PFES	Payment for Forest Environmental Services
PKK	Worker's Party of Kurdistan (Kurdish; Partiya Karkerên Kurdistanê)
PNRH	National Water Resources Plan (Portuguese)
PWA	Phillip Williams and Associates
RAP	Resettlement Action Plan
RECOFTC	The Centre for People and Forests
REDD	Reduce Emissions from Deforestation and Forest Degradation
REDD+	Reduce Emissions from Deforestation and Forest Degradation in Developing Countries
RF	Rockefeller Foundation
RoG	Republic of Ghana
RSDSC	Red Sea-Dead Sea Conveyance
SAARC	South Asian Association for Regional Cooperation
SDGs	Sustainable Development Goals
SEDS	Socio-Economic Development Strategy
SFEs	State Forestry Enterprises
SHP	Soil Health Program
SIWI	Stockholm International Water Institute
SLBE	State Level Boomerang Effect
SME	Small and Medium Enterprise
SOE	State Owned Enterprise
SRI	System of Rice Intensification
SSA	Sub-Saharan Africa
STIMC	Save the Tigris and Iraqi Marshes Campaign
TAK	Kurdistan Freedom Hawks
TBGWB	Transboundary Ground Water Basin
TGD	Three Gorges Dam
TWR	Turkish Water Report
UFP	Universidade Federal do Para
UN	United Nations
UNCED	United Nations Conference on Environment and Development
UNEP	United Nations Environment Programme
UNESCO	United Nations Educational Scientific and Cultural Organization
UNFCC	United Nations Framework on Climate Change
UNFCCC	United Nations Framework Convention on Climate Change
UNHCR	United Nations High Commissioner for Refugees
UN-REDD	United Nations Programme on Reducing Emissions from Deforestation and Forest Degradation

US	United States
USAID	United States Agency for International Development
WASH	Water Sanitation and Hygiene
WCD	World Commission on Dams
WEC	World Energy Council
WEF	World Economic Forum
WHO	World Health Organization
WWAP	World Water Assessment Programme
WWF	World Wildlife Fund

1 The boomerang effect

Overview and implications for climate governance

Larry A. Swatuk, Lars Wirkus, Florian Krampe,
Bejoy K. Thomas and Luis Paulo Batista da Silva

Introduction

The world is ramping up its actions towards combating human-induced climate change. Through the COP processes, national governments have committed to a wide variety of mitigation and adaptation actions through their NDCs (Nationally Determined Contributions) to emissions reduction. While actions are to be primarily taken at national level, it is made clear that 'parties may use internationally transferred mitigation outcomes to achieve NDCs' (Article 6 of the Paris Agreement). This opens a path for such things as carbon-market development, biofuels production and other forms of green energy and green economy development. How these will all be measured and evaluated is rather opaque. It is clear, however, that developing countries, particularly those most vulnerable to climate change, are least able to design, implement, monitor and evaluate climate action interventions. Billions of dollars are to be made available for these actions through mechanisms such as the Green Climate Fund (GCF); and billions more are likely to be generated through largely artificially devised carbon markets.

While the narrative of 'global threat leading to collective action' seems straight forward, the actual landscape of global climate governance is fragmented and fraught with contradictions, conflicts and conundrums (Widerberg and Pattberg, 2016). Moreover, the 'crisis' aspect of the narrative encourages states to scramble around for examples of 'good practice', relabelling and marketing existing development interventions in the name of climate change adaptation and mitigation. Forests have suddenly become not a living entity both intrinsically valuable and instrumentally valuable to animals and humans alike, but a category for meeting emissions targets. Agriculture has become both threat and opportunity in relation to both adaptation and mitigation: particular agricultural practices are regarded as greenhouse-gas-emission–heavy; others are seen to be climate friendly. While the transformation from the former to the latter is seen as a potential multi-purpose win, rare is the question ever asked 'What does this mean for social stability?' Whereas many countries have built into their national climate strategies the need for a 'pro-poor approach', the reality is that 'agriculture' – like 'forests' – stands to become a category, not a socio-cultural process embedded within a particular geographical landscape, yielding value beyond just food and/or profit.

While national actions plans and programmes are intended to yield multiple benefits to humans and nature – i.e. 'climate security' (Boas and Rothe, 2016; Gemmene *et al.*, 2014; Dabelko *et al.*, 2013; Barnett, 2007; Barnett and Adger, 2007) – failure to adapt to or mitigate the hypothesized effects of climate change is forecast to result in dramatic social and environmental instability, including mass migration and resource-related violent conflict (German Advisory Council on Climate Change, 2007). While effects will be felt unevenly both within and among states and regions, the overall outcome is anticipated to be negative (Dalby, 2013; Bernauer *et al.*, 2010; Wisner *et al.*, 2003; Bohle *et al.*, 1994). However, in the rush to action there is also the danger of generating unanticipated and unintended negative impacts. These negative impacts have drawn different labels such as 'maladaptation' and 'back-draft', and different scholarly and policy-oriented communities have begun to theorize ways to avoid them (Dabelko *et al.*, 2013). For this to happen, an integrated, transdisciplinary approach is necessary. A recent policy paper commissioned for the G7 argues that climate change adaptation/mitigation actions require a 'conflict-sensitive approach' that integrates both climate and socio-economic-political vulnerabilities (Ruttinger *et al.*, 2015). This recommendation stems from the view that climate change will overburden weak states, may also destabilize strong states (cf. Moran, 2011), and have negative regional impacts (Conca, 2001).

In this chapter we introduce 'the boomerang effect', defined here as the emergence of largely unanticipated and unintended negative consequences of climate change adaptation and mitigation policies and programmes on domestic non-state actors that result in negative feedbacks on the state. By 'state', we mean government actors taking decisions as representatives of the state. The chapter has three objectives. First, to contribute directly to theory by articulating a framework for analyzing one particular aspect of maladaptation, that is, 'the boomerang effect'. Second, to present an overview of the chapters in this collection reflecting on the real and potential unanticipated and unintended negative effects at the local level (local-level side effects – LLSEs) and at the state level (state-level boomerang effects – SLBEs). Third, to draw lessons from the cases for research, policy and practice. Each case study engaged (more or less directly) with four primary research questions:

- What are the (social/economic/ecological/political) drivers behind a particular development or climate intervention?
- What was the decision-making process that led to this specific climate action or development intervention?
- What are the LLSEs (social/economic/ecological/political) of the action and are any of these unintended and/or unanticipated and negative in consequence?
- What are the boomerang effects felt by the state?

Each chapter also addressed a policy-oriented question: Recognizing that there will always be uneven outcomes and maladaptive practices, what are better processes to minimize negative impacts? In this chapter, we aggregate the findings and distil the lessons to be learned. In this way, we hope to assist policymakers in avoiding both LLSEs and SLBEs as climate action–oriented development interventions are rolled out.

Climate action and 'boomerang effects': towards a framework of analysis

The impacts of climate security actions involving land-use change are being studied intensively (see, e.g. CCMCC research programme www.nwo.nl; Magnan *et al.*, 2016) and are being integrated into a variety of literatures that, for our purposes here, may be considered together under the term 'climate security' (Swain, 2015; Scheffran *et al.*, 2012; Smit and Wandel, 2006; O'Brien and Leichenko, 2000; Graeger, 1996). Within this broad church, there is also broad disagreement, as methods, episto-methodologies, ontologies, analytical frameworks and ideological perspectives vary widely (Bavinck, Mostert and Pellgrini, 2014; Gleditsch, 2012; Schnurr and Swatuk, 2012; Detraz and Betsill, 2009; Matthew *et al.*, 2009; Grothmann and Patt, 2005). The result is a highly fragmented field of study (Bräutigam and Zhang, 2013; Anseeuw *et al.*, 2012). Our research project, of which this edited collection is a part, aims to contribute directly to theory by developing a framework for analysing the boomerang effect.

To elaborate somewhat: Deriving from state actor climate change adaptation or mitigation policies and programmes, the implementation of these climate interventions (through state or state-authorized private actors) often has unanticipated and unintended negative social, political, economic and ecological effects that impact on local communities on various spatial and temporal scales. These impacts in turn negatively feed back to the state on multiple levels (e.g. local, regional, national), at various scales (e.g. watershed, forest, landscape, ecosystem), with numerous impacts (e.g. political economic instability, social unrest and violence), thus undermining climate security.

This definition builds on similar framings such as the Wilson Center's idea of 'backdraft' (Dabelko *et al.*, 2013), and the extensive work on 'maladaptation' (Barnett and O'Neill, 2013; Barnett and O'Neill, 2010; McCarthy *et al.*, 2001; Scheraga and Grambsch, 1998), defined by the IPCC in its AR5-WGII report as 'actions that may lead to increased risk of adverse climate related outcomes, increased vulnerability to climate change, or diminished welfare, now or in the future ' (quoted in Field *et al.*, 2014).

In this chapter, we specifically articulate local-level and state-level impacts and discern their interrelationship, particularly the negative – or 'boomerang' – effects felt by the state (see Figure 1.1) thereby refining the ability to differentiate among 'maladaptive' impacts. We differentiate these complementary effects in terms of (i) local-level side effects and (ii) state-level boomerang effects. In Figure 1.1 it can be seen that climate action–oriented policies initiated by

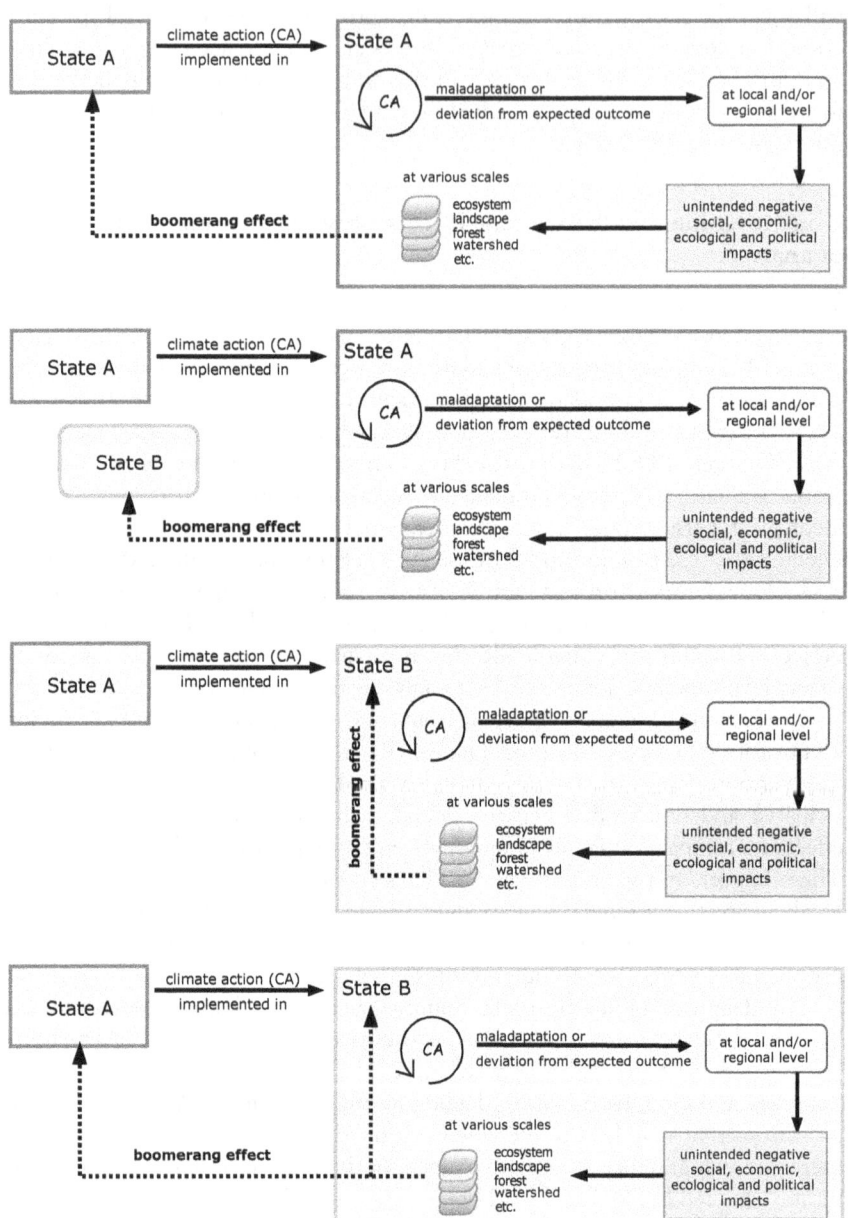

Figure 1.1 The boomerang effect.

one state actor can result in boomerang effects along (at least) four different pathways. All four cases see climate actions initiate local-level side effects in the target country, but it is also possible for regional local-level side effects to occur (as shown in Chapter 4 in this collection), possibly leading to boomerang effects in either the state that initiated the action or/and in the target state (as is happening with land and water grabbing across the African landscape).

In our collection, local-level side effects are delineated along blue-water (hydraulic infrastructure such as dams and canals) and green-water (forest conservation, biofuels development) pathways – and may be discerned through empirically demonstrable social, economic, political and ecological impacts. These indicators measure the local-level impacts that derive as side effects of implemented climate change adaptation and mitigation programmes and policies. State-level boomerang effects may be discerned through empirically demonstrable impacts manifesting as threats to economic stability, state authority and/or ecological sustainability. This typology differs from that developed by Magnan (2014) (in his framework for avoiding environmental, sociological and economic maladaptation) in its specific focus on local-level impacts and state-level impacts resulting from explicitly defined and traceable climate security actions. Boomerang effect indicators measure governance-level impacts that are the feedback loop that derive from the local-level side effects and impacts. These indicators assess three critical threats to the state that emerge on multiple levels (e.g. local, regional, national) and at various scales (e.g. watershed, forest, landscape, ecosystem) (see Table 1.1 below). By disaggregating real and potential effects in this way, we aim to assist government actors to devise more socially equitable, environmentally sustainable and economically efficient policies and programmes.

Green water and blue water

Many claims regarding the positive or negative outcomes of climate change actions rest on a series of over-generalizations regarding land (e.g. that it is degraded, over-utilized or under-utilized), water (e.g. wasted, limited, poorly utilized) and people (e.g. 'too many'; lacking skills or motivation/incentives). In our view, better information must be made available to inform decisions at various stages and of various stakeholders. Important is the distinction between green and blue water. Green water is water that is utilized by plants from the soil directly following rainfall. Productive green water is defined as that which transpires through a plant, creating biomass. Unproductive green water is defined as rainfall that evaporates directly back to the atmosphere. Blue water is that which is available as run-off after rainfall. It takes the form of surface water (rivers, lakes, streams, impounded behind dam walls) and readily accessible subsurface water, i.e. groundwater (through borehole/well technology) (Falkenmark and Rockstrom, 2004). Swatuk et al. (2015) further refine unproductive green water into a 'socio-ecological unproductive pathway', meaning water that is productively used by plants that either (i) are destructive of the local ecosystem

Table 1.1 Local-level side effects and boomerang effects from case studies

Cases		Local-level side effects				State-level boomerang effects		
Country	Action	Social	Economic	Political	Ecological	Economic stability	State authority	Ecological sustainability
Chapter 2: TGD	Dam	1.5 million displaced	Mixed (positive and negative)	Social protest	Dramatic change to aquatic and terrestrial environment	High costs associated with relocation, environmental rehabilitation, etc	None discernible	Government argues multiple benefits from flood management
Chapter 3: Guarani	Aquifer Management Agreement (not activated)	Numerous discrete urban and rural water-supply problems	Business as usual with uneven benefits	Site-specific protests	Worries about aquifer health from poor waste management (cities, farming, mining)	Costs 'normalized'	None discernible	Soil degradation, deforestation, urban sprawl all treated as routine within government departments
Chapter 4: Farakka	Water diversion project (symbol of technological mastery of man over nature)	Small-scale farmers displaced in Bangladesh	GDP impacts from agricultural challenges	Social protest	Dramatic change to dry season ecology	Numerous costs for Bangladesh	None discernible but persistent poor inter-state relations	Dramatic challenges to all in GMB basin
Chapter 5: Belo Monte	Dam	Displacement	Mixed	Social protest	Dramatic changes to aquatic and terrestrial environment	Internationalization of issue has numerous knock-on effects in Brazil	Corruption scandals lead to government change	State announces no further dams in the region

Chapter	Type							
Chapter 6: UN-REDD	Forest conservation and rehabilitation	Mixed	Mixed	Mixed	Mixed	None discernible	None discernible	Government argues multiple benefits from REDD activities
Chapter 7: Ilisu	Dam	Displacement	None discernible	Social protest	Dramatic change to aquatic and terrestrial environment	High costs associated with internationalization of the issue (e.g. loss of investors)	Persistent political instability in region	Downstream basin states claim significant change to ecology
Chapter 8: Jordan	Desalination and other technological innovations	Persistent social instability in relation to Palestine	Mixed	Social protest	Living 'beyond the water barrier'	Mixed	Regional relations securitized	Local and basin-wide evidence of resource degradation
Chapter 9: Ghana	Drought-resistant agriculture	Intra-household and intra-community conflict	Mixed	Small-scale disputes	Cash-crop focus risks long-term chance to ecological sustainability	None discernible	None discernible	Scaling-up of approach may negatively affect regional ecosystem
Chapter 10: Gigel Gibe III	Dam	Health of local people affected through malaria; downstream people affected by loss of access	Increase in ill-health impacts livelihoods; loss of access to water impacts livelihoods	Small-scale disputes	Dramatic ecosystem change	None discernible	Regional relations strained	Research suggests downstream ecosystems being seriously altered

(e.g. alien or invasive species); and/or (ii) ultimately benefit only a few users (e.g. privately owned sugar cane plantations exploiting land and labour for profits accruing to the few).

This more nuanced understanding of green/blue water will provide an important means for assessing the likelihood of success for intended mitigation/ adaptation actions in relation to short, medium and long-term socio/political-ecological system resilience. For example, biofuels expansion into lands deficient in green water may result in compound (environmental/social/ political) negative effects while yielding limited (environmental/economic) benefits.

There is an important class-related distinction to green/blue water. Across the Global South, the best land – i.e. with robust soil-water holding capability and, if necessary, ready access to supplemental blue-water irrigation – has been captured by large-scale commercial agriculture, relegating smallholder peasants to land that is more difficult to farm. These are overwhelmingly green-water–dependent landscapes, meaning that farmers depend on rainfall for crop production. It is these marginal lands that now are being targeted for expansion within the framework of climate change adaptation/mitigation activities. The potential for social upheaval leading to spontaneous and/or organized violence, therefore, is very real.

Green-water pathway

It is generally argued that climate change is impacting hydrological cycles leading to increasing difficulties for regions and countries to meet their food security needs. In consequence, states and private sector actors have entered into a wide variety of land-use agreements designed to ensure win–win outcomes: food, water, livelihood and state security through better crop choice and/or food production in areas best suited for production. This is an adaptation pathway.

At the same time, countries have agreed to limits on carbon emissions. One means for many coal-dependent countries to meet these limits is to encourage increased production of crops for biofuels. Another is to conserve forest lands – either their own or in another state through the purchase of carbon credits – for carbon sequestration. Many states, private sector actors, and non-governmental organizations are actively engaged in these practices through REDD and REDD+ (Gallemore and Jespersen, 2016; Lund *et al.*, 2016; Abbott, 2012). These are mitigation pathways. In this collection, Chapters 6 (Vietnam) and 9 (Ghana) specifically deal with green-water pathways.

Blue-water pathway

It is generally agreed that climate change will lead to more extreme events. It also will lead to more water in some places and less water in others; and to widely fluctuating hydrological cycles that will be increasingly unpredictable. To ensure water security for human activities, therefore, this unpredictability must be dealt with through infrastructure development – what Conca (2006) calls, 'damming, diverting and draining'. This is primarily an adaptation pathway, though multi-purpose hydraulic infrastructure often claims mitigation elements as well, where, for example, hydropower displaces thermal-power as a primary means of electricity generation. In this collection, Chapters 2 (China), 4 (India–Bangladesh), 5 (Brazil), 7 (Turkey) and 10 (Ethiopia) showcase the negative local-level side effects and (real and potential) boomerang effects from big infrastructure projects.

Mixed approaches

Largely depending on scale, blue-water and green-water interventions often bleed into each other. For example, forests make use of both green and blue water. As peasant farmers alter the landscape for agricultural purposes, they remain largely dependent upon rainfall for crop production. This is green water. But, as UN-REDD programmes encourage forest rehabilitation, planted trees will ultimately and predominantly tap groundwater, which is blue water. The distinction is important, as impacts at the scale of the intervention and at the scale of the watershed will be different. Similarly, primarily blue-water interventions such as multi-purpose dams make it possible for significant extension of agriculture (via irrigation) into formerly green-water-dominant areas. Put differently, the application of technology and capital make it possible for powerful actors to displace subsistence farmers in the interests of cash-crop production (i.e. 'land grabbing' and 'water grabbing').

In our collection, in contrast to the other cases, there are two that present quite unique challenges for policymakers regarding climate change and environmentally sustainable, socially equitable and economically efficient resource use: large-scale transboundary aquifers and transboundary river basins shared by highly antagonistic states. In the case of the former, the Guarani Aquifer, shared among Brazil, Paraguay, Argentina and Uruguay is treated in Chapter 3. In the case of the latter, the Jordan River Basin, shared among Syria, Jordan, Lebanon, Israel and Palestine is the focus of Chapter 8. In both cases, these large basins show the importance of collective management in the face of changing climates and socio-economic demographics. However, state actors are resistant to anything other than state-based approaches serving 'national interests'. Irrespective of blue-water (dams, canals, pipelines, large-scale irrigation) or green-water (expansion of rainfed agriculture) activities in these basins, states continue to rely on either technological innovation and trade (Chapter 8) or a case-by-case approach (Chapter 3) to problem solving. While the local-level side effects are numerous and varied, the boomerang effects appear either manageable or latent, so giving to policymakers an illusion of security.

Case studies

While it is generally agreed that climate actions of different types will result in unintended negative consequences, impacting people at different scales in different ways around the world (see www.newsecuritybeat.org/category/blog-columns/backdraft-podcast/ for numerous examples and insights), our project aims to disaggregate these effects so as to better inform both our understanding of their dynamics and to assist state, civil society and private-sector actors to target their interventions so as to minimize the social, economic, environmental and political costs. We distinguish between LLSEs and SLBEs in order to turn a spotlight on the 'boomerang effect' in the belief that improved policy-making is more likely when state actors are made aware through empirical evidence of the direct and potentially serious negative impacts that derive from their resource use decisions (see Table 1.1 for a disaggregation of these effects). The papers in this collection, except for Chapter 9, derive from desk studies overseen by Swatuk. The goal was to assemble information enabling the authors of this introductory chapter to better understand the impacts of climate action with a view towards developing a sustained research project. In this way, we hope to contribute to the emerging literature discussed above. As we developed the case studies it became clear that disaggregating the local-level effects from those at the state level was important. On the one hand, it helped explain why obvious policy missteps were tolerable in some instances but not in others. On the other hand, it helped us discern insights and approaches to better policy-making. We return to this last point near the end of the chapter.

Chapter 2 focuses on China's Three Gorges Dam. This massive exercise in hydraulic infrastructure raises important questions regarding how those at the receiving end of state-directed development – climate action–oriented or otherwise – share in the project's design, direction, implementation and benefits. The global track record on dams and development does not offer much hope in this regard. After a short lull in construction before and after the convening of the World Commission on Dams (see WCD, 2000), big dam-building exercises are underway in earnest across the Global South, much of it facilitated by China. The climate adaptation and mitigation narrative undergirds many of these exercises, all of which are being championed as necessary to ensure water and energy security in a climate change–affected world. As shown in Table 1.1, the local-level side effects are numerous and dramatic, especially in relation to displaced people and flooded landscapes. However, the state-level boomerang effects are few, containable and/or tolerable, suggesting that the Chinese government is confident with its approach to development in the face of current and anticipated climate challenges.

Over the last 10–15 years, increased focus has been given to the world's groundwater resources. Due to changing demographics, less reliable rainfall forecasting and changing market demands, human settlements and farm enterprises of all sizes are increasingly turning to groundwater. In consequence, great stress is being placed on this resource, with the Ogallala Aquifer being an important case in point. The Guarani Aquifer is the subject of Chapter 3. In the mid-2000s the four states that share the aquifer – Brazil, Argentina, Paraguay, Uruguay – endeavoured to arrive at a transboundary groundwater basin management arrangement. This was fostered by a great deal of international support.

As with many other international agreements, however, the 2010 Guarani Aquifer Agreement has foundered on the rocks of national ratification. It is only a small exaggeration to say that it is 'business as usual' across the basin. While not directly related to commitments made at COP21, the case does raise questions about the ways and means of fostering necessary cooperation when parties perceive the impacts to be tolerable and manageable, discreetly felt and only tangentially linked to state-level negative effects.

Chapter 4 focuses on the Farakka Barrage, a water-diversion project initiated by India several decades ago but which serves as a symbol of the folly of state-led actions taken in the 'national interest' to 'bend Mother Nature' to the will of 'Man'. We use the noun 'man' deliberately here, as it is primarily male engineers and politicians who continue with an unflagging faith in the combined application of technology and capital to 'solve' problems related to resource access, use and management. Tony Allan (2003) describes this as 'the hydraulic mission', most typical in the West during the early-to-mid part of the twentieth century, but, as illustrated in this collection, very typical of large parts of the Global South today. The Farakka Barrage is a 'river-training' project, and forms one part of India's long-standing grand scheme of linking 'its' rivers in a manageable grid so that areas of shortage may find the necessary supply to satisfy their varied demands. If four decades of global water-governance has taught us anything, it is the value of a basin-wide approach, involving all relevant stakeholders, and mobilizing all relevant forms of knowledge. Many of the problems the states of the Ganges–Megna–Brahmaputra (GMB) basin face derive from their fragmented, nationalist approaches to resource planning. Successful adaptation and mitigation activities require collective engagement and agreement, yet this case suggests that India has no intention of moving in this direction. The LLSEs and SLBEs resulting from such an approach are many and costly: mass migration; diminished agricultural production stemming from significant changes to the ecology; reinforcement of poor political relations between India and Bangladesh. This is hardly a firm foundation upon which to take collective climate action in the GMB basin.

While Bangladesh has had marginal success in internationalizing India's 'river-training' intentions and interventions, groups most negatively affected by Brazil's Belo Monte Dam project have had much more success, but also with significant LLSEs. As shown in Chapter 5, local actors have partnered with global civil society groups to press the Brazilian government to address their numerous serious concerns. International civil society pressure has had demonstrable SLBEs. For example, exposing the numerous LLSEs to world scrutiny and helping to blow the whistle on government corruption in the construction industry not only helped bring down the Rousseff government, but led the Temer government in January 2018 to declare that no further mega-dams would be built in the Amazon region (see https://news.mongabay.com/2018/01/brazil-announces-end-to-amazon-mega-dam-building-policy/). Granted, the dam is operational, but government was forced to alter its design so as to become more socially and environmentally responsible. Indeed, the government of Brazil continues to tout the Belo Monte as a symbol of its commitment to clean and renewable energy.

In Chapter 6, the authors examine the case of UN-REDD in Lam Dong province, Vietnam. Vietnam is extensively involved in REDD and REDD+ programmes, mainly due to the immense pressures rural populations (through sheer numbers) and big agriculture (through land conversion) are putting on forest resources. The case study is most interesting in that it highlights the mixed outcomes at local and national level. In important ways, the introduction of internationally supported programmes and projects such as REDD serve to either reinforce or alter existing social relations at the local level. The economic opportunities created by REDD create new scrambles for resources that exacerbate existing social tensions. While SLBEs are not pronounced (REDD goals contrast with government development policy, so creating some tension), LLSEs are mixed as winners and losers are created under the new REDD regime.

The Ilısu Dam development project, the subject of Chapter 7, reflects the other dam cases in this collection. As with the Belo Monte and TGD cases, the Ilısu will result in the displacement of tens of thousands of local people by submerging towns and villages under the reservoir. Similar to the Belo Monte, the winners and losers from the scheme are not only divided in terms of geography (local and rural losers; urban winners located far from the site of the development project) but in terms of ethnicity. Tribal groups in Brazil are mirrored in Turkey by the Kurds. Each group is able to mobilize global networks of support, so creating SLBEs in both cases: start-up delays resulting in economic loss, loss of political capital regionally and globally. So effective have been local groups in mobilizing against the government that the project has been delayed several times due to creditors pulling out of the project. Like China, however, the Turkish government is determined to fund the project with or without international financial support.

As with the Guarani Aquifer, so with the Jordan River Basin (JRB), the subject of Chapter 8: states in the basin are not interested in pursuing collective approaches to resource management. Whereas Brazil is the upstream basin hegemon in the Guarani, Israel is the downstream basin hegemon in the Jordan. Neither sees a need to collaborate with others at this time. Unlike the Guarani with its unratified GAA, JRB states have not attempted basin-focused collaboration. Granted the political situations are completely different, with the JRB being securitized and all questions of resource access, use and management being filtered through a high political lens. What does this mean for water and related resource security in the JRB in the era of climate change? As shown in the chapter, all states continue with the fiction that nationalist approaches will lead to resource security, with Israel in particular holding firm to the high-modern belief that technological innovation holds the key. In this way, Israel resembles India where money and technology make 'go-it-alone' policies, programmes and practices seem practical and sufficient.

Chapter 9 focuses on northern Ghana and presents a different sort of picture in relation to LLSEs. Here the real and potential negative effects manifest within households (between men and women) and within and between communities (also in a highly gendered context). Government programmes in Ghana are being designed to facilitate climate-resistant agriculture. These are being rolled out with the assistance of state, civil society and private sector actors within Ghana, across sub-Saharan Africa and the wider world. As shown in this chapter, however, projects that focus on staple crop production favour men and discriminate against women. At the same time, the gendered nature of agricultural production reflects not only social relations, but the way livelihood practices are embedded within ecosystems. Shifting towards scaling up certain forms of production may in the long run have the opposite of the intended effect: jeopardizing, instead of supporting, long-term ecological sustainability. There are no discernible SLBEs, so providing little feedback to the state regarding the appropriateness of the programme.

Similar to the Ghana case study, Chapter 10 focuses on the gendered health- and livelihood-related LLSEs resulting from dam building in Ethiopia. The SLBEs resulting from the Gilgel Gibe III Dam on the Omo River are well known in relation to poor inter-state relations between Ethiopia (touting energy for development) and Kenya (concerned about the degradation of Lake Turkana downstream). Some of the LLSEs are also obvious: as with other dams, the displacement of people from their homes and lands as well as dramatic ecosystem change that stresses livelihoods. Less well-known are the health impacts related to the creation of large bodies of standing water. Canals and dams in particular environments act as disease vectors, in this case malaria. The gendered aspect of the threat relates to women's role as managers of household water and the fact that they are most often left behind when men migrate to find work in cities. These LLSEs go largely unnoticed except by health and community workers who are left with 'triage-oriented' approaches to personal, household and community security, e.g. highly localized and generally poorly funded WASH (water, sanitation and health) or anti-malarial (bed net) programmes. As with the rural and remote peoples in the other dam cases, the ill-health of communities located around the Gilgel Gibe III Dam appears to be an expense the state is willing to bear.

Lessons for research, policy, planning and practice

As shown by the case studies in this collection, development interventions – whether climate action–specific or not – create winners and losers. As the global governance architecture shapes itself around large-scale climate actions for planetary-wide impacts, the papers here serve as cautionary tales of the danger of failing to carefully consider the LLSEs and SLBEs that result from the rush to 'solutions'. The dam cases – Chapters 2 (TGD), 5 (Belo Monte), 7 (Ilısu), 10 (Gilgel Gibe III) – are particularly instructive in this regard, as the LLSEs and SLBEs are numerous and very serious. If one includes the Farakka Barrage (Chapter 4) case as well, there is strong empirical evidence showing that where money, power and interests coalesce, projects are pushed forward despite the evident LLSEs. In addition, the tolerability to state actors of the (social/economic/ecological) costs to be borne by local communities varies directly with the distance between the site of the intervention and the primary beneficiaries (national, regional, global). This bodes ill for those resident at sites of intervention (e.g. carbon sequestration; renewable energy through wind, solar or hydro) meant to benefit 'the planet' (<2°C). At the same time, our case studies reveal that where there are SLBEs, the tolerability of these effects varies directly with the degree of difficulty in handling them. The Communist Party of China (CPC) seems unassailable in this regard. In contrast, the democratic character – however weakly embedded the democracy – of Brazil seems to provide space for social movements and civil society organizations to press for concessions that will in fact be attended to. This observation will come as cold comfort to those who have suffered at the hands of the dam builders and their supporters. It is also clear that where the intervention has been framed as necessary for state security (Chapters 4, 7 and 8) there is almost no room for reconsideration of the scope and form of the proposed project.

The evidence presented here is not all 'doom and gloom'. There is evidence that better outcomes may result when local, national and global interests are in alignment. Yet even here, the so-called 'local' rarely if ever presents as a unified entity, and activities undertaken – despite the best of intentions – will create winners and losers and sometimes reinforce existing animosities (Chapter 6). Based on the evidence drawn from the cases, recommendations for better climate action and development practice may be divided into four categories: (i) participation, planning, policy and institutions; (ii) perspective; (iii) alternatives; (iv) framings.

Participation, planning, policy and institutions

Many of the chapters argue for institutional and policy reform. Both Chapters 4 (Farakka) and 8 (Jordan) argue in support of a basin-wide management structure. Chapter 3 (Guarani) argues for ratification and operationalization of the GAA. Most chapters argue for meaningful participation in the planning process. Chapter 2 (TGD) authors suggest that policy-making will improve through more democratic and transparent processes. The authors also argue that the media and civil society could play important watch-dog roles. In Chapter 6 (Vietnam), the authors also argue in support of more transparent planning and policy-making processes, involving local communities, indigenous people and minority groups in meaningful ways.

Perspective

Several chapters critique existing approaches to planning and policy, suggesting that decision makers are hampered by their commitment to outmoded, 'high-modern' perspectives. What is needed is a commitment to integrated planning, so: integrated water resources management (IWRM) (Chapter 4); economic/ecological/social integration (Chapter 2); a holistic perspective (Chapter 5);

a regional approach (Chapters 3, 4, 8 and 10) and a commitment to complementary activities (Chapter 9) steeped in different ways of knowing and knowledge gathering (Chapters 2, 6, 7, 9 and 10).

Alternatives

Such an approach to planning and policy-making would reveal, in the opinion of many of the authors, alternatives to 'inflexible infrastructure' with exorbitant sunk costs at a time when flexibility and adaptability are central to 'climate-proofing' the national and regional environment. For example, in Chapters 2, 5, 7 and 10, the authors argue for localized solutions that should be multiple and at small scale.

Framings

It is clear from the cases presented here that how a project is presented – to 'beneficiaries', to funders – matters a great deal. In some cases, dominant discourses obstruct actors' abilities to move beyond the status quo. The Jordan River Basin case (Chapter 8) shows this most clearly, but it is evident throughout the 'big infrastructure' chapters. Often times the dominant framing forces those who are unsatisfied with the status quo to take the opposing position: i.e. dam/no dam; canal/no canal. Thus the parties remain locked in a contest presented as zero-sum. The challenge is to rearticulate the opportunities available to dominant decision makers. This is an important lesson for all who are interested in climate change mitigation and adaptation policy and practice. Just because it is 'good for the planet', does not mean that it is fair and equitable and hence the right thing to do.

At the outset we articulated five questions regarding project planning and implementation. They are summarized in Table 1.2 below. In short, it can be seen that the drivers behind the projects described above are arrayed around the 'usual suspects' – development, profit, sustainability, social benefit – but only tangentially speak to power projection by the state, or in the case of Brazil, India and Israel, the demonstration of hegemony in a particular basin.

Analysis of the decision-making processes reveals an absence of meaningful participation by those most seriously affected by the project. Thus, there are numerous LLSEs that are probably unintended but possibly anticipated and deemed tolerable by the state. While there were many traceable SLBEs, they all appeared to be not serious enough to derail a state from its intended course of action. At best, some alteration to a project – e.g. the Belo Monte Dam – was evident.

As we have also shown here, there are better ways of making plans, policy and carrying out projects. Whether these will be mainstreamed into climate action–oriented projects is not clear to us.

Table 1.2 Matrix of impacts and opportunities

Research question	Observation
Drivers behind a particular development or climate intervention?	– development – profit – sustainability – projection of power by the state – social benefit
Decision-making process that led to this specific climate action or development intervention?	– top down – in some cases process was more open and inclusive – mostly lacking in transparency and accountability
LLSEs (social/economic/ecological/ political) of the action and are any of these unintended and/or unanticipated and negative in consequence?	– many social, economic, ecological and political impacts across the different cases – majority unintended but probably easily foreseen and deemed tolerable by government
Boomerang effects felt by the state?	– yes, many economic, social, political and ecological at different scales and intensities – all regarded as ultimately tolerable by state
Better processes to minimize negative impacts?	– more participatory and open planning processes – institutional and policy reform – alternatives to the preferred option considered and weighed by all affected by the project/programme – appropriately framed to affect positions, interests and needs of all stakeholders

Conclusion

For those at the frontline of environmental change, improving livelihoods and alleviating poverty are the appropriate frameworks for dealing with complex vulnerabilities, including environmental insecurity (Deligiannis, 2012). From the evidence amassed here, it is doubtful that development and/or climate action–oriented policies and plans put people before profit, or align what's good for the planet so that it is good for the people, particularly those most seriously affected by the planned intervention. We are saying that as global development and climate governance continues with top-heavy approaches to managing both 'sustainable development' and 'two degrees', it is especially important to organize appropriate adaptation and mitigation responses from the grassroots and to then reach up and out to all relevant stakeholders. No other approach will yield sustainable co-benefits.

References

Abbott, K. W. (2012). The transnational regime complex for climate change. *Environment and Planning* C: *Government and Policy*, 30, 571–590.

Allan, J.A. (2003). IWRM/IWRAM: *A New Sanctioned Discourse?* Occasional paper 50 (April). SOAS Water Issues Study Group. London: School of Oriental and African Studies/King's College London.

Anseeuw, W., Boche, M., Breu, T., Giger, M., Lay, J., Messerli, P., and K. Nolte (2012). *Transnational Land Deals for Agriculture in the Global South. Analytical Report based on the Land Matrix Database.* Bern/Montpellier/Hamburg: CDE/CIRAD/GIGA.

Barnett, J. (2007). Environmental security and peace. *Journal of Human Security*, 3(1): 4–16.

Barnett, J., and Adger, W. N. (2007). Climate change, human security and violent conflict. *Political Geography*, 26(6), 639–655.

Barnett, J., and O'Neill, S. (2010). Maladaptation. *Global Environmental Change – Human and Policy Dimensions*, 20, 211–213.

Barnett, J., and O'Neill, S. J. (2013). Minimising the risk of maladaptation: a framework for analysis. In: J. P. Palutikof *et al.*, eds, *Climate Adaptation Futures*. Hoboken, NJ: Wiley-Blackwell, pp. 87–94.

Bavinck, M., Pellegrini, L., and Mostert, E., eds (2014). *Conflict on Natural Resources in the Global South: Conceptual Approaches.* Boca Raton, FL: CRC Press.

Bernauer, T., Kalbhenn, A., Koubi, V., and Ruoff, G. (2010). Climate change, economic growth, and conflict. Paper presented at the conference Climate Change and Security, Trondheim, Norway 21–24 June.

Boas, I., and Rothe, D. (2016). From conflict to resilience? Explaining recent changes in climate security discourse and practice. *Environmental Politics*, 25(4), 613–632.

Bohle, H. G., Downing, T., and Watts, M. J. (1994). Climate change and social vulnerability: Toward a sociology and geography of food insecurity. *Global Environmental Change*, 4(1), 37–48.

Bräutigam, D., and Zhang, H. (2013). Green dreams: myth and reality in China's agricultural investment in Africa. *Third World Quarterly*, 34(9), 1676–1696.

Conca, K. (2006). *Governing Water.* Cambridge, MA: MIT Press.

Conca, K. (2001). Environmental cooperation and international peace. In P. Diehl and N. P. Gleditsch (eds.), *Environmental Conflict* (Boulder and Oxford: Westview Press), pp. 225–247.

Dabelko, G., Herzer, L., Null, S., Parker, M., and Sticklor, R. (2013). *Backdraft: The Conflict Potential of Climate Change Adaptation and Mitigation.* Washington, DC: Environmental Change and Security Program, Wilson Center.

Dalby, S. (2013). Biopolitics and climate security in the Anthropocene. *Geoforum*, 49, 184–192.

Deligiannis, T. (2012). The evolution of environment-conflict research: Toward a livelihood framework. *Global Environmental Politics* 12(1), 78–100.

Detraz, N., and Betsill, M. M. (2009). Climate change and environmental security: for whom the discourse shifts. *International Studies Perspectives*, 10(3), 303–320.

Falkenmark, M., and Rockstrom, J. (2004). *Balancing Water for Humans and Nature.* London: Earthscan.

Field, C. B., Barros, V. R., Dokken, D. J., Mach, K. J., Mastrandrea, M. D., Bilir, T. E., Chatterjee, M., Ebi, K. L., Estrada, Y. O., Genova, R. C., Girma, B., Kissel, E. S, Levy, A. N., MacCracken, S., Mastrandrea, P. R., and White, L. L., eds (2014).

Climate Change 2014: Impacts, Adaptation and Vulnerability. Contribution of Working Group II to the Fifth Assessment Report of the Intergovernmental Panel on Climate Change. Cambridge: Cambridge University Press.

Gallemore, C., and K. Jespersen, (2016). Transnational markets for sustainable development governance: The case of REDD+. *World Development*, 86, 79–94.

Gemenne, F., Barnett, J., Adger, W. N., and Dabelko, G. (2014). Climate and security: evidence, emerging risks, and a new agenda. *Climatic Change*, 123(1), 1–9.

German Advisory Council on Global Change (2007). *Climate Change as a Security Risk.* London and Sterling: Earthscan.

Gleditsch, N. P. (2012). Whither the weather? Climate change and conflict. *Journal of Peace Research*, 49(1), 3–9.

Graeger, N. (1996). Environmental security? *Journal of Peace Research*, 33(1), 109–116.

Grothmann, T., and A. Patt (2005). Adaptive capacity and human cognition: The process of individual adaptation to climate change. *Global Environmental Change* 15(3), 199–213.

Hsiang, S. M., and Burke, M. (2013). Climate, conflict, and social stability: what does the evidence say? *Climatic Change*, 123(1), 39–55.

Khagram, S., and Ali, S. (2006). Environment and security. *Annual Review of Environment and Resources*, 31(1), 395–411.

Lund, J. F., Sungusia, E., Mabele, M. B., and Scheba, A. (2016). Promising change, delivering continuity: REDD+ as conservation fad. *World Development* 89, 124–139.

Magnan, A. (2014). Avoiding maladaptation to climate change: towards guiding principles. *Sapiens*, 7(1): 1–11.

Magnan, A. K., Schipper, A. L. F., Burkett, M., Bharwani, S., Burton, I., Eriksen, S., Gemenne, F., Schaar, J., and Ziervogel, G. (2016). Addressing the risk of maladaptation to climate change. *Wiley Interdisciplinary Reviews: Climate Change*, 7(5), 646–665.

Matthew, R. A., Barnett, J., McDonald, B., and O'Brien, K. L. (2009). *Global Environmental Change and Human Security.* Boston, MA: MIT Press.

McCarthy, J. J., Canziani, O. F., Leary, N. A., Dokken, D. J., and White, K. S., eds (2001). *Climate Change 2001: Impacts, Adaptation and Vulnerability. Contribution of Working Group II to the Third Assessment Report of the Intergovernmental Panel on Climate Change.* Cambridge: Cambridge University Press.

Moran, D., ed. (2011). *Climate Change and National Security.* Washington, DC: Georgetown University Press.

O'Brien, K. L., and Leichenko, R. M. (2000). Double exposure: assessing the impacts of climate change within the context of economic globalization. *Global Environmental Change*, 10(3), 221–232.

Ruttinger, L., Smith, D., Stang, G., Tänzler, D., and Vivekananda, J. (2015). *A New Climate for Peace.* Berlin: Adelphi, International Alert, Woodrow Wilson International Center for Scholars, European Union Institute for Security Studies.

Scheffran, J., Brzoska, M., Brauch, H. G., Link, P. M., and Schilling, J., eds (2012). *Climate Change, Human Security and Violent Conflict.* Berlin, Heidelberg: Springer Science and Business Media.

Scheraga, J. D. and A. E. Grambsch (1998). Risks, opportunities and adaptation to climate change. *Climate Research* 10, pp. 85–95.

Schnurr, M. A., and Swatuk, L. A., eds (2012). *Environmental Change, Natural Resources and Social Conflict: Towards Critical Environmental Security.* London: Palgrave.

Smit, B., and Wandel, J. (2006). Adaptation, adaptive capacity and vulnerability. *Global Environmental Change*, 16(3), 282–292.

Swain, A. (2015). Water and post-conflict peacebuilding. *Hydrological Sciences Journal*, 61(7), 1313–1322. http://doi.org/10.1080/02626667.2015.1081390

Swatuk, L. A., McMorris, M., Leung, C., and Zu, Y. (2015). Seeing 'Invisible Water': challenging conceptions of water for food, agriculture and human security. *Canadian Journal of Development Studies*, 36(1), 24–37.

Widerberg, O., and Pattberg, P. (2016). Accountability challenges in the transnational regime complex for climate change. *Review of Policy Research*, 34(1), 68–87.

Wisner, B., Blaikie, P., Cannon, T., and Davis, I. (2003). *At Risk: Natural Hazards, People's Vulnerability and Disasters* (2nd edn). London and New York: Routledge.

2 Nothing's always perfect

The boomerang effect of Three Gorges Dam

Bojian Zhang, Chieh Cheng, Yuye Li, Zhe Zhang and Larry Swatuk

Introduction

Hydropower development has gained ground from the climate change adaptation and mitigation discourse. This chapter focuses on the Three Gorges Dam (TGD) and its tremendous impacts on different areas in China. We begin by introducing factors influencing hydropower development in China. In section 3, the TGD is discussed along with a brief summary of its primary purposes and foreseen benefits. Section 4 explores in detail the local-level side effects resulting from this huge project, including natural biology, environmental problems and cultural heritage, and two more profound influences, economic and social aspects. The final section draws some relevant lessons from this study for hydropower development in China and beyond.

Hydropower status quo in China

China is currently experiencing unprecedented economic development, and is now the second largest economy in the world just behind the US. Coal is a driving force behind this economic rise, accounting for 65 per cent of China's total energy supply. This makes China the number one emitter of CO_2 and SO_2 worldwide, and almost every big city in China is facing air pollution issues at different levels, seriously impacting human health. So China is practically forced to seek for alternative, renewable and clean energy sources to gradually take the place of traditional fossil fuel, like coal.

Hydropower seems like a good energy source: it does not pollute air because no fossil fuels are burned; it is renewable because it uses the earth's water cycle to generate electricity; the cost of producing hydropower is relatively stable, hardly affected by the increasing costs of fossil fuel; and it is a source of power with high flexibility since stations can ramp up and down very quickly to adapt to different power demand (Worldwatch Institute, 2012). Data collected by the World Energy Council (2016: 7) shows that '[i]n recent years China has taken centre stage for hydropower capacity, accounting for 25 per cent of global installed capacity in 2015, far ahead of USA (8.4%), Brazil (7.6%) and Canada (6.5%)'. After China's twelfth five-year plan (2011–2015), the state council announced the goal that boosting hydropower to a 15 per cent share of the country's primary energy consumption was basically realized, and it is estimated that this share is still climbing. 'Total capacity in China is expected to reach 350 GW of pure hydropower and 70 GW of pumped storage by 2020' (WEC, 2015: 7 quoting data from China's twelfth five-year plan).

Although hydropower is the most cost-effective method of generating energy from a renewable source, should hydropower really be considered as a clean power source? The simplest answer is 'sometimes'. Here this study takes Three Gorges Dam (TGD) as an example, to argue that this mega project while beneficial for national development in terms of electricity generation, will in the longer term produce far-reaching negative environmental, economic and social local-level side effects. Whether these will translate into state-level boomerang effects is uncertain. In the words of Stone (2008), '[i]n China, public debate about the dam's dark side is muted.' The relatively non-conflictful process of relocating more than one million people while dramatically altering the natural and built environment suggests great leeway for government to continue to undertake big infrastructure projects of sometimes dubious merit.

Three Gorges Dam

Brief overview

China already has half of the world's largest dams, including the greatest one – Three Gorges Dam (TGD). The dam is located on the Yangtze River at 30"49'23" N latitude and 111°0'12" E longitude. The Yangtze River Basin comprises 20 per cent of China's total land area, and is home to some 400 million people. Major cities located along its main stem are Chongking, Wuhan, Nanjing and Shanghai (see http://factsanddetails.com/china/cat15/sub99/item460.html for details). While Chongking is upstream of the dam, Wuhan, Nanjing and Shanghai are located downstream from the dam. As the largest hydroelectric station in the world, the dam cost $59 billion and took 15 years to build. The dam wall is 2.3 km long and 185 metres tall. The reservoir is 660 km long and has flooded an area of some 632 km², equivalent to the size of Switzerland, drowning more than 1,000 towns and villages and displacing more than one million people (Handwerk, 2006; IRN, 2003). There are 32 main turbines with a total capacity of 22,500 MW equivalent to electricity generated by 20 coal-fired or 18 nuclear power stations (see www.power-technology.com/projects/gorges/ and www.reuters.com/article/us-china-weather-power/in-drastic-move-chinas-top-hydropower-plants-slash-capacity-idUSKBN19P0D6 for details). The controversial dam has numerous benefits, including flood control, enhanced river trade, and emissions reduction, but significant costs as well (see below).

History of TGD

The dam has been the dream of Chinese leaders for more than 80 years, including Sun Yat-Sen, Mao Zedong and Deng Xiaoping. In fact, the initiative of building a dam on Three Gorges was originally envisioned by Sun Yat-sen in 1919, who stated that a dam capable of generating 30 million horsepower (22 GW) could be built on the Yangtze River. In 1944, a US engineer, John L. Savage, surveyed this area and drew up a dam proposal for the 'Yangtze River Project', but due to the Chinese Civil War, this project was halted (Travel Guide China, n.d.). After 1949, the Communist Party of China (CPC) came to power, and the 1954 Yangtze River floods, which killed an estimated 30,000 people, doubtless reinforced Mao Zedong's inclination to build this dam (Mufson, 1997). In 1992 the National People's Congress approved the dam. Construction started on 14 December 1994 and was completed in 2015 (see https://en.wikipedia.org/wiki/Three_Gorges_Dam for a useful overview). Three Gorges Dam was, to borrow words from former prime minister Li Peng, 'a symbol of the superiority of the socialist system'; at the same time it is also a symbol of human's capacity to remake nature (Mauch *et al.*, 2006: 8).

The dam derives its name from its location in the famous Yangtze Gorges region of China. The dam is located nearest to Xiling Gorge (66 km long) and downstream of Qutang Gorge (8 km long) and Wuxia Gorge (42 km long) (see https://whc.unesco.org/en/tentativelists/1623/). According to Zheng (2015: 143),

> the Yangtze Three Gorges scenic area crosses Hubei Province and Chongqing Municipality. As a result of this administrative division, Wuxia Gorge and Qutang Gorge fall within Chongqing, while Xiling Gorge, a section of Yangtze River to Badong, and two dams, the Gezhou Dam and now the Three Gorges Dam, fall within Hubei Province.

The dam itself is located at Sandouping, Yichang County, Hubei Province (Heming and Rees, 2000).

Main purposes of TGD

Essentially, there are three primary purposes for the TGD: power generation, flooding control and navigation. In terms of power generation, the operation of TGD power station can generate 84.7 TWh (billion kilowatt hours) power every year, which can greatly relieve the electricity tension in the middle and eastern parts of China. At the same time, it has a huge contribution to make to reduce the emission of carbon dioxide and other greenhouse gas, an amount equal to that produced by burning 50 million tons of coal (Stone, 2008). So at this point, the positive effect of this hydropower station to environmental protection is affirmed.

Regarding flood control, the Yangtze is famous for its floods, earning the name 'river dragon' in China. Frequent floods have routinely caused problems throughout the Yangtze River catchment and the Dongting Lake area and the dam is designed to prevent such events in future (Hayashi *et al.*, 2008). If severe floods happen, the project can efficiently control the flow of water from upstream and the big dam can hold the water, which buys time for transferring personnel to avoid loss of life.

In terms of navigation, due to its location in the middle of Yangtze River, the TGD dramatically promotes river traffic. The increasing width and depth of the Yangtze River allows ships of 10,000 tons to sail directly from Chongqing to Wuhan and further to Shanghai, so greatly reducing transport costs (Worldwatch Institute, 2012). By 2013 the system had already exceeded the cargo targets set for 2030. According to Huifeng (2014),

> [t]he tonnage of shipping using the five-tier lock exceeded 100 million for the first time in 2011, more than five times the level before the dam was constructed. It reached a record high of 100.06 million tonnes last year, representing 40,848 cargo vessels and 2,461 passenger ships, or 300,000 tonnes per day.

Economic benefits brought by the TGD

As suggested above, there are significant economic benefits to be derived from the TGD project. Initially, the TGD project attracted a massive amount of investments. The total static investment of the TGD is 90 billion yuan. Among all the investments, construction investment accounts for around 50 billion yuan, and the resettlement investment accounts for 40 billion yuan. On the other hand, the total dynamic investment accounts for over 200 billion yuan (Cook and Murray, 2003). The impact of enhanced electricity generation on GDP is significant, as are the benefits of flood control to the prevention of disruptions to economic and social life.

Significant direct benefits derive from such a huge project. A high demand for building materials benefited cement, concrete and steel industries, among many others. Indeed, countless numbers of companies were eager to get involved in the project, sensing the generation of considerable profits, and, with this experience under their belts, the ability to win tenders and participate in China's national and global dam-building spree (Webber, 2011; Swatuk, 2018).

As a new man-made wonder of the world, the world's largest hydropower project draws vast numbers of tourists every year, which brings considerable tourism revenue to the two administrative regions in which the three gorges lie. Data in Zheng (2015: 146) illustrate the dramatic rise in international tourist arrivals to the cities of Yichang and Chongqing between 1990 and 2011: from 48,600 in Yichang in 1990 and 62,600 in Chongqing to 248,600 and 1.864 million respectively in 2011.

Local-level side effects

In addition to these demonstrable benefits, the negative local-level side effects caused by this project are extensive and likely to worsen over time. The chapter now turns to a brief discussion of the environmental, economic and social aspects of these.

Natural biology

The construction of the TGD and associated reservoir has negatively impacted flora and fauna both upstream and downstream in various ways (Wu *et al.*, 2004; Xu *et al.*, 2013). For example, the Yangtze's four major carp species (bighead, black, grass and silver) spawn when water levels rise during the summer monsoon rains (Stone, 2008). However, the dam has altered seasonal variations in water levels below the dam. As a consequence, there have been declines recorded in carp eggs and larvae downstream. Other affected species include the Chinese River Dolphin (known as the baiji, or white-fin), South China tiger, Chinese Sturgeon, Giant Panda and so on. Lei (1998) pointed out that the 'Chinese River Dolphin is the living fossil of a 30-million-year-old species', and by the turn of the twenty-first century the number remaining is fewer than 100. Lei reports that the water released from upstream washes away the river islets downstream, so negatively impacting fish habitat, jeopardizing dolphins' main food source. In addition, the TGD has led to a dramatic increase in shipping, with ship collision causing accidental death of the river dolphin (Lei 1998).

Environmental problems

As shown in Table 2.1 below, the major environmental problems stemming from the TGD are numerous: e.g. decline in water quality, reservoir sedimentation, soil and riverbed erosion, seismic activity, geological hazards and flood.

Water quality

The water quality in the reservoir has dropped notably since 2003. Increased pollutant concentration and slow water flow caused by impoundment in some bays of the reservoir have led to eutrophication (Xu *et al.*, 2013), which is the (over)abundance of nutrients in an aquatic system. In some cases, such as Xiangxi Bay – a tributary bay of the Three Gorges reservoir – algal blooms have resulted (Yang *et al.*, 2010). If severe enough, localized waters will become hypoxic (oxygen poor) or, in extreme cases, anoxic (fully depleted of oxygen) leading to severe stress and/or death of aquatic life.

Table 2.1 An overview of the major environmental problems of the TGD

Environmental issue	Evidence	Stakeholder	Interest	Outcome
Water quality (Eutrophication)	1 Eutrophication and algal bloom have become notable in many bays of the reservoir and some tributaries since 2003	Fishermen	Better catch of fish	The number of fish they harvest is dropping each year
Sedimentation and riverbank erosion	2 The mean of sediments is 142 million tonnes per year, which equals 40% of the estimated 355 million tonnes per year in the Environmental Impact Statement for the Yangtze Three Gorges Project (EIS) Report. 3 The average of annual erosion rate from October 2002 to October 2010 is 108.8 million m3, and it is higher than the average of annual erosion rate from 1975 to 2002	1 Residents 2 Shipping companies	1 Better quality and stable quantity of water 2 Safety	1 Since the problem of sedimentation, the reservoir might not provide good quality and stable quantity of water 2 It is not safe to navigate in a reservoir if the sedimentation problem becomes too severe; the salient of sediment might cause ship collision. 3 Erosion may cause geological hazards, i.e. landslide
Seismic activity	Reservoir-induced earthquake shows a high frequency and low intensity pattern in EIS Report	1 Residents	A stable living environment	High frequency earthquakes may cause landslide, earthquake and landslide are both threats to residents' safety
Downstream flooding risk	In July 2014, there was a reservoir-caused flood in Chongqing	1 Farmers 2 Residents	1 A stable living area 2 Lands for farming	The flood forced 4,000 residents to evacuate from their home, it also killed 11 people, and 27 people were missing.

Sedimentation and erosion

There are several reasons for both riverbank and riverbed erosion. Flooding behind the impoundment has destabilized shorelines leading to increased erosion along the banks of rivers and an increase in sedimentation behind the dam wall. It is estimated that over the first 100 years of its lifespan, the TGD will retain more than 44 per cent of the total sediment load deriving from the river's upper reaches (Yang *et al.*, 2010: 1). At the same time, timed release of less turbid water from the dam boosts the scouring ability of water downstream. Decreased sediments downstream will lead to a vicious cycle whereby without curbs like sediments the scouring ability of the river's flow with be enhanced. This will dramatically alter the health and character of the aquatic environment (Yang *et al.*, 2010). Mining and farming activities along the river have also increased riverbank erosion while simultaneously increasing nutrient and heavy-metal–laden silt loads so jeopardizing water quality and estuarine health.

Sedimentation determines the capacity of storing water and the lifespan of a reservoir. Because the capacity of the reservoir is limited by the extent of its design, if sediment occupies too much capacity of the reservoir, there will be less space for water. When a reservoir is full of sediment and is unable to be cleaned, then this reservoir has reached its lifespan. Given the nature of economic activity (e.g. urban and agricultural expansion) and natural rainfall patterns (with heavy seasonal rains), the sediments washing into the Yangtze behind the impoundment are extensive, so limiting the life of the reservoir as well as its capacity for both flood control and electricity generation.

Seismic activity and geological hazards

One of people's greatest fears is that the TGD is built on an earthquake zone, and the land condition is fragile. The massive amount of water produces a lot of pressure, when the pressure below the earth heightens, it is likely to induce earthquake so negatively affecting immediate and adjacent areas (Edmonds, 1992).

Flood

The dam is supposed to control the flood problem in Yangtze region; however, heavy rain, i.e. storms and plum rains, would fill the reservoir, and there is no other choice but to release the surplus water, which causes floods in downstream. The dam has protected downstream areas from flooding for most of the time, but when storm or plum rain season comes, the downstream area may well lose the protection from the dam. To this point in time, it seems that the TGD has done its job in relation to flood control. However, the extensive flooding across the lower reaches of the Yangtze during 2016s heavy rainy season suggests two things. One, the dam offers a false sense of security for those downstream, for less flooding does not mean no flooding. Second, it seems clear that reduced sediment-flows downstream are having a negative impact on the flood resilience of cities and towns along the banks of the river. This is clearly an unintended consequence of a dam designed to lower the risk of flooding (Jing, 2016).

Cultural heritage

There are so many areas being flooded due to the TGD that the damages to sites of valuable and significant historical heritage is a problem. Some sites record the nation's history back to Neolithic times (Li et al., 2016). The TGD has brought grave challenges to the protection and exploitation of cultural relic. At the same time, due to poor administrative work and insufficient funds, heritage conservation planning approval is slow, making this work much harder (Hvistend-ahl, 2008). By the end of 2000, the land that was inundated was estimated to contain around 1,282 cultural heritages (Ponseti and López-Pujol, 2006).

Economic problems

The environmental challenges sparked off a series of economic problems, for example, fishery and farming. From the fishery aspect, there were an estimated 300 species of fish in the Yangtze River. However, pollution, eutrophication and dramatic changes to the character of the river have combined to lower the fish catch in Yangtze region. In a 2017 news report, *China Daily* stated that the annual Yangtze fish catch was now below 100,000 tons. This has led the government to ban fishing in all nature reserves along the river in an effort to help restore the health of the ecosystem (see www.chinadaily.com.cn/china/2017–12/01/content_35148214.htm; also, Stone, 2008).

In order to deal with water problems, the government had to budget 22.8 billion yuan for the 'Water Pollution Prevention Plan for the Three Gorges Reservoir and the Upper Reaches of the Yangzi River', for enhancing the water quality of the reservoir and trying to solve the problem of eutrophication (Ministry of Ecology and Environment, 2011. Every year, the Chinese government has to spend a large amount of money on addressing environmental issues, which increases the burden on state finances (Cook and Murray, 2003). While the Chinese government should be one of the biggest beneficiaries from the TGD project, there is a clear economic boomerang effect making the government pay the price for breaking the balance between ecological systems.

From a farming aspect, construction of the dam sacrificed some of the arable lands; it is indicated that the TGD flooded approximately 34,000 hectares of farmlands. Among these farmlands, 50 per cent is rice fields, and 10 per cent is forests (Jackson and Sleigh, 2000). As a result, Xu et al. (2013) claim that the TGD contributes to a grain shortage of 150,000 tons every year for the whole China. Proponents of the dam emphasize comparative value:

> It was estimated that for every one billion kilowatt hours in annual output, inundation of 28.35 hectares of land was needed for the TGD, compared

with an average of 900 hectares of land for the other 31 large and medium size dams currently under construction.

(Zhu, 1996)

However, what proponents fail to consider is the possibility that arable lands – whatever the size – are more valuable than hydropower. Moreover, the opportunity cost of letting arable lands lie under the water level is pretty high; people in China have to use 7 per cent of the world's arable lands to feed 25 per cent of the world's population (Jackson and Sleigh, 2000). In sum, it is a trade-off between electricity and arable lands, and seemingly, choosing electricity over arable lands may have higher economic loss.

Social problems

While numerous social issues have arisen due to the negative local-level side effects discussed above, without doubt the most serious problem concerns relocation. As of June 2008, the Chinese government had relocated 1.24 million residents, as 13 cities, 140 towns and 1,350 villages were flooded either wholly or partly by the filling of the reservoir (International Rivers, 2016).

Numerous relocatees have been forced from or been involuntarily displaced from their homes. And at some times, police have used excessive force to curb small and peaceful protests, and organizers have been beaten and imprisoned. They all can be counted as human right abuses (Boyle *et al.* 1998).

Emigration resettlement is also a social-economic behaviour. Proper resettlement can be beneficial for the region's development; but once the resettlement of immigrants fails, it can threaten thesecurity and stability of society. The area surrounding the Yangtze River is one of the most populous areas in the world, and the construction of the dam has caused an immense impact in terms of flushing out this area of its inhabitants. People have to migrate to other places and find new jobs. There are more than 15 million people in the TGD region. The population in this region have low levels of education, 25 per cent of them are illiterates, and the annual GDP of the TGD area in 1992 was $190 US dollars. It is reported that 40 per cent of relocatees are farmers, but only 60 per cent of them received farm land as compensation after resettlement (McLaughlin, 2011). Thus, the other 40 per cent of farmers lost not merely their employment and income but their livelihoods. As a result, they have to move to urban areas to look for work. As collective and widespread unemployment problems lead to economic loss, so economic loss can negatively affect public security. Every year, there are roughly 180,000 labour-related protests in China. According to a report from Probe International, the Chinese government now spends more money on domestic security than its military budget (McLaughlin, 2011). Complicating the issue of resettlement are state practices associated with individual/household compensation and regional investment. Each of these issues are marred with reports of corruption, reflecting in part the distance those displaced by the TGD are from the centres of economic and political power

(Hvistendahl, 2008). In the longer term, these flash points may trigger violence directed at the state.

Lessons learned from the TGD

There are two kinds of lessons that we can learn from the TGD – pre-construction and post-construction. In terms of pre-construction lessons, it seems obvious that prior to launching major building projects, an improved process of policy-making is the precondition of making right choices. A variety of alternative plans should be taken into consideration, including smaller-scale dams and other forms of renewable energy. It is clear in the case of the TGD that the question of whether or not to build was never asked. Given the fact that water is (social, economic, political) power, the only questions asked regarding the TGD were 'When?', 'How?' and 'How much will it cost?' Following completion of these large-scale projects, there tends to be a number of unexpected side effects, most of which are severe and long-term, so monitoring and post-construction governance are necessary in order to minimize negative local-level side effects and avoid state-level boomerang effects. From our point of view, the TGD project seems to reflect the central state's belief in its ability to control economically and/or politically derived social unrest and manage – through relentless technical tinkering – environmental stressors. Thus, as the world moves forward into another dam-building era, what are the necessary take-aways from our case study?

Improved policy-making at a higher level of democracy and transparency

The brief history of the Three Gorges project demonstrates weak existence of democracy (Edmonds, 1992). First, the approval by the National People's Congress (NPC) was just a 'rubber-stamp'. The role of the congress was always considered to be 'to automatically approve all that was put before it' (Edmonds, 1992). Second, China's leaders were a vital determinant of such a mega-scale project. Tiananmen Square protests strengthened the power of Li Peng and Jiang Zemin, who were both known as supporters of the project (Cook and Murray, 2003). Third, the mass of the people in China could hardly have access to the comprehensive information of the Three Gorges project. Since the 1991 flood that was used as 'a case in point' to exaggerate the function of flood control, the voices heard in the media had all been in favour of it (Edmonds, 1992). So most people, even some representatives of the NPC, could hardly gain a thorough understanding of the project, with available information being limited. Edmonds (1992) reports that there was a representative from Tianjin Municipal who abstained from voting due to not realizing 'why the congress had been asked to vote on a major project for which they had no technical expertise'. Some engineers who spoke out against the project were even imprisoned. Far from today's development mantra of 'stakeholder involvement', the TGD is simply a product of the will of government.

China's political power is concentrating in the hand of the CPC, so another persistent problem remains – the lack of transparency and the corruption that comes along with it. It was reported that, in 2000, there were nearly 100 officials convicted of corruption while working on the TGD. Official corruption is a big threat to the construction of any mega-project, not only in China but around the world. The safety standards in construction were easily influenced by local officials and builders, resulting in potential cracks in the structure of the dam (Fennell, 1999). Several bridge collapses upstream were identified as a result of suspected faulty construction by a contractor who bribed local officials.

The story of the TGD has taught the Chinese government a painful lesson that if it does not allow appropriate consultation and public participation on major issues of collective interest, there can be a huge price to pay (International Rivers, 2016). In order to make the decision-making process more democratic under the current social and political system in China, the following aspects should be taken into account. First, comprehensive and impartial information about proposals presented at the NPC should be available to all the representatives, and in some cases, it is necessary for experts working in a certain domain to analyze solid and sober information about the project and make neutral reports, objectively evaluate their advantages and disadvantages, taking all aspects into consideration. Second, the Chinese People's Political Consultative Congress (CPPCC), an advisory institution where democratic parties participate actively in the deliberation and administration of national affairs and the supervision of policy implementation, are supposed to play a key role in the process of policy-making through submitting independent reports after thorough investigation. Although the Three Gorges project was finally approved, the CPPCC reported that the dam would produce more damage than benefit, which led to a fierce debate about the project. In our view, the CPPCC should have a more formal role beyond simply being advisory. Finally, legislation should be passed to limit government intervention into mass media, provide for independence of the press and protect journalists' rights. Mass media are considered as an effective means of monitoring the process of decision-making and administration, providing relevant information for the public promptly. Democracy in policy-making means equally valuing different voices, especially the voice of the public. Mass media is an important vehicle for people's voices to be heard.

Making a careful decision: an alternative plan for a single giant dam with multiple purposes

The Chinese government's approach to water resources management reflects an abiding belief in the power of man over nature. The TGD is simply one of a number of ongoing and planned big infrastructure projects being undertaken within China (e.g. the South-to-North Diversion Project) and by Chinese companies around the world (Swatuk, 2018). Clearly, policymakers need to rebalance the relationship between ecological, economic and social aspects, reducing the political influence of usual policy enforcers. Critics of TGD argued that a viable alternative to the single giant dam was a series of smaller scale projects along the Yangtze River and on its tributaries upstream and downstream (Edmonds, 1992). These, they argued, would have had positive environmental and socio-economic consequences, while minimizing the wide array of negative local-level side effects that have manifested as a result of the project. More importantly, in our view, is the opportunity cost of sunk investment into massive, inflexible infrastructure. In an era of climate change-induced uncertainty, flexibility and adaptive capacity will be the hallmarks of sustainability – not giant showpiece projects.

If there had been an integrated project consisting of a series of small-scale dams along the Yangtze River and on its tributaries, the situation would have been much better: the short construction time for each smaller dam could have allowed for compensating for costs quicker, without creating a great economic burden on China; there would have been far fewer migrants in need of resettlement and much less loss of the fertile farmland that can hardly be made up by the new fields in resettlement areas. There would have been fewer concerns about the increasing possibility of earthquake, which lies in the additional pressure placed upon the bottom of the huge reservoir, and no concerns about the existence of a common military target (Edmonds, 1992).

Developing a wide variety of renewable energy resources

Compared with traditional energy, like coal, oil and natural gas, hydropower is comprehensively considered as the most efficient energy form, but in the meanwhile, the 'blood and tears' of history keeps reminding us that hydropower is still a double-edged sword, which can not only benefit humankind, but also destroy us and our posterity. Therefore, factors, like environmental pollution, social impacts, project security, power plant location and so on, must be taken into account as part of the multitude of considerations in building new dams, which means both planners and investors must consider sustainability.

At the same time, some truly eco-friendly and renewable energy resources, such as solar, wind, tidal and other newly emerging forms of electricity generation, need to get enough funding for development, and they all have very optimistic prospects, especially in countries with high energy consumption, like China. So, government should actively develop new and renewable energy, based on the specific situation within a certain area, with the focus placed on the installation of equipment and technical improvement, diversifying the energy supply mix, facilitating the energy use, and ensuring more secure, stable, eco-friendly and sustainable energy supplies.

In China, hydropower has played a dominant role in renewable energy supply, accounting for about 78 per cent of the whole in 2010, but its growth rate is much lower than those of wind, solar and biomass energy (Zhang et al., 2012). According to the research by Hu and Cheng (2013), thanks to a long coastline and spacious territory, China has great potential wind energy, with wind power the second largest source of renewable electricity generation. It is estimated that the annual amount of electricity generated by wind could reach 6.96 trillion kWh, approaching the demand in 2030, at the price of about 0.062 – 0.095 $/kWh. In addition, China has a vast land mass that is endowed with rich biomass resources and widely distributed geothermal resources, so biomass and geothermal energy would have a bright future.

Problem-solving mechanism on post-construction environmental and socio-economic issues

As for environmental issues, if construction of such large hydro projects is completed, it means that relevant problems cannot easily be avoided. So, to seek to address the environmental challenges caused by the Three Gorges Dam, based on the research by Xu *et al.* (2013), it might be helpful to establish the long-term environmental monitoring systems and joint operations with other large projects in the upstream areas. Additionally, it is suggested that large-scale projects, like the TGD, should involve a strategic environmental assessment from a broader perspective. Regular assessments focusing on specific environmental issues are also necessary. Given that prevention of negative local-level side effects and boomerang effects is impossible after construction, the earlier the emerging problems are monitored, the more possible it would be to minimize the damage on the natural and built environment, and on humans and nature alike.

The current geological and meteorological systems, which comprise a considerable number of branches located in the entire Yangtze River basin, lay a foundation for the network of environmental monitoring, with minimized demand for extra funding and technicians. Mass media could play an active role in these monitoring systems through reporting new environmental issues, which could be affected by local governments, so less intervention on mass media benefits the process of diminishing damage on the natural environment as well.

In terms of managing socio-economic affairs, effective and democratic governance functions as a key determinant, and local governments should be in cooperation with local communities and consultative individuals or institutions, like some impartial social scientists, to deal with social issues, like resettlement of migrants, and economic issues, such as compensating for the loss of farmlands and fisheries of those migrants. Governance requires more relevant stakeholders to be involved, so it is necessary to build intermediate organizations, where government officials and local community leaders could share each other's opinions and coordinate their interests, and social scientists could pass on useful advice. It is worth mentioning that all the stakeholders enjoy equal status, meaning that local governments are in a series of problem-solving partnerships with local communities and third-party experts, instead of just giving previous administrative guidance. Such practices contribute to strong democracy in policy-making. For example, compensation for the loss of agricultural fields and fisheries and funding for resettlement could be set up in cooperation with affected farmers and fishermen, civil society organizations and so on, instead of being determined solely by government officials.

Actually, the better strategy is to avoid building the TGD. As long as the dam is there, people have to pay for all the negative effects on the social and natural environment. Even if post-construction governance is good enough, it can never remedy catastrophic results caused by a bad decision.

Conclusion

In conclusion, as the biggest hydroelectric dam in the world, the TGD is responsible for great benefits to the people of the Yangtze River Valley, and the whole of China. However, the negative local-level side effects and state-level (largely economic) boomerang effects are significant across environmental economic and social landscapes. Lessons must be learned from the TGD in order to recover and protect the environment and ecological systems around the dam, to reduce the negative impacts on economic development, to help those people who lost their land, jobs and income, and essentially, to avoid the occurrence of similar projects.

Despite the disadvantages, hydropower has broad market prospects in China. As governments are adopting policies to mitigate climate change, renewable energies like hydropower will be key winners. By ensuring stakeholders' meaningful participation, continuous innovation in technology, better governance and management practices and more critical and reflective planning, many of the mistakes of the TGD can be avoided. Whether this can happen in China is doubtful, but the lessons hold for other societies facing similar challenges.

References

Boyle, P., Halfacree, K., and Robinson, V. (1998). *Exploring Contemporary Migration*. London: Longman.

Cook, I. G., and Murray, G. (2003). *Green China: Seeking Ecological Alternatives*. London: Routledge.

Edmonds, R. L. (1992). The Sanxia (Three Gorges) Project: the environmental argument surrounding China's super dam. *Global Ecology and Biogeography Letters*, 105–125.

Fennell, T. (1999). Power struggle: Criticism, corruption and costs plague the Canada-aided Three Gorges dam project. *Maclean's*, 112(33), 36–37.

Handwerk, B. (2006). China's Three Gorges Dam, by the numbers. National Geographic News (9 June). Available at: https://news.nationalgeographic.com/news/2006/06/060609-gorges-dam.html. Accessed 15 May 2018.

Hayashi, S., Murakami, S., Xu, K. Q., and Watanabe, M. (2008). Effect of the Three Gorges Dam Project on flood control in the Dongting Lake area, China, in a 1998-type flood. *Journal of Hydro-environment Research*, 2(3), 148–163.

Heming, L. and Rees, P. (2000). Population displacement in the Three Gorges reservoir area of the Yangtze River, Central China: Relocation policies and migrant views. *International Journal of Population Geography*, 6, 439–462.

Hu, Y., and Cheng, H. (2013). Development and bottlenecks of renewable electricity generation in China: A critical review. *Environmental science and technology*, 47(7), 3044–3056.

Huifeng, H. (2014). Three Gorges Dam exceeds cargo target set for 2030. *South China Morning Post* (24 May). Available at: www.scmp.com/news/china/article/1519185/three-gorges-dam-exceeds-cargo-target-set-2030. Accessed 31 March 2018.

Hvistendahl, M. (2008). China's Three Gorges Dam: An environmental catastrophe? *Scientific American*. Available at: www.scientificamerican.com/article/chinas-three-gorges-dam-disaster/. Accessed 31 March 2018.

International Rivers (2016). Three Gorges Dam. Available at: www.internationalrivers.org/campaigns/three-gorges-dam. Accessed 31 March 2018.

International Rivers Network (IRN) (2003). Human rights dammed off at Three Gorges. Available at: www.internationalrivers.org/sites/default/files/attached-files/3gcolor.pdf. Accessed 31 March 2018.

Jackson, S., and Sleigh, A. (2000). Resettlement for China's Three Gorges Dam: socio-economic impact and institutional tensions. *Communist and Post-Communist Studies*, 33(2), 223–241.

Jing, L. (2016). Flood of doubts: sceptical public questions Three Gorge's Dam's capacity to stop disasters. *South China Morning Post* (17 July). Available at: www.scmp.com/news/china/article/1991055/flood-doubts-sceptical-public-questions-three-gorges-dams-capacity-stop. Accessed 31 March 2018.

Lei, X. (1998). CHINA: taking the eco-pulse of a giant. *Science*, 280(5360), 25–25.

Li, J., Fang, H., and Underhill, A. P. (2016). The history of perception and protection of cultural heritage in China. In: A. P. Underhill and L. C. Salazar, eds, *Finding Solutions for Protecting and Sharing Archaeological Heritage Resources*. New York: Springer, pp. 1–16.

Mauch, C., Stoltzfus, N., and Weiner, D. R. (2006). *Shades of Green: Environmental Activism Around the World*. Lanham, MD: Rowman and Littlefield.

McLaughlin, K., (2011), For development, China moves millions: Global Post. *Probe International*. Available at: http://journal.probeinternational.org/2011/09/21/for-development-china-moves-millions-globalpost/. Accessed 31 March 2018.

Ministry of Ecology and Environment (2011). *The National Twelfth Five-year Plan for Environmental Protection (2011-2015)*. Beijing: The People's Republic of China.

Mufson, S. (1997). The Yangtze Dam: Feat or folly? *Washington Post* (9 November): p. AO1.

Ponseti, M., and López-Pujol, J. (2006). The Three Gorges Dam project in China: history and consequences. *HMiC: història moderna i contemporània*, (4), 151–188.

Stone, R. (2008). Three Gorges Dam: Into the unknown. *Science*, 321(5889), 628–632.

Swatuk, L. A. (2018). The land-water-food-energy nexus: green and blue water dynamics in contemporary Africa–Asia Relations. In: P. M. A. R de Medeiros Carvalho, D. Arase and S. Cornelissen, eds, *Routledge Handbook of Africa-Asia Relations*. London and New York: Routledge, pp. 386–405.

Travel Guide China (n.d.). *Benefits from Three Gorges Project*. Available at: www.travelchinaguide.com/attraction/hubei/yichang/three-gorges-dam-project.htm. Accessed 31 March 2018.

Webber, M. (2011). The political economy of the Three Gorges Project. *Geographical Research*, 50(2), 154–165

Wu, J., Huang, J., Han, X., Gao, X., He, F., Jiang, M., … and Shen, Z. (2004). The three gorges dam: an ecological perspective. *Frontiers in Ecology and the Environment*, 2(5), 241–248.

World Energy Council (WEC) (2016). World Energy resources: Hydropower 2016. Available at: www.worldenergy.org/wp-content/uploads/2017/03/WEResources_ Hydropower_2016.pdf. Accessed 31 March 2018.

Worldwatch Institute. (2012). Use and capacity of global hydropower increases. Available at: www.worldwatch.org/use-and-capacity-global-hydropower-increases. Accessed 31 March 2018.

Xu, X., Tan, Y., and Yang, G. (2013). Environmental impact assessments of the Three Gorges Project in China: Issues and interventions. *Earth-Science Reviews*, 124, 115–125.

Yang, Z.-J., Liu, D.-F., DaoBin, J. I., and Xiao, S.-B. (2010). Influence of the impounding process of the Three Gorges Reservoir up to water level 172.5 m on water eutrophication in the Xiangxi Bay. *Science China* 53(4), 1114–1125.

Zhang, X., Chang, S., and Eric, M. (2012). Renewable energy in China: An integrated technology and policy perspective. *Energy Policy*, 51, 1–6.

Zheng, Q. (2015). Crisis Management, Tourism and the Three Gorges Dam, China. PhD Thesis. University of Central Lancashire, UK. Available at: http://clok.uclan.ac. uk/11808/1/Zheng%20Qiying%20Final%20e-Thesis%20%28Master%20Copy%29. pdf. Accessed 31 March 2018.

3 Transboundary groundwater governance and management

The case of the Guarani Aquifer – Brazil, Argentina, Paraguay, Uruguay

Kadra Rayale, Kaylia Little, Mary Crawford and Larry Swatuk

Introduction

Groundwater basins are used for their freshwater reserves and as such, are essential to human livelihoods. Over time, humanity's increased reliance on groundwater as a chief source of fresh water is due in large part to the growth of industry, agriculture and the global population (Eckstein and Eckstein, 2005). These factors coupled with the real effects of climate change have stimulated discussions on the importance of maintaining the world's freshwater reserves. In a transboundary water basin setting, this resource transcends administrative or political boundaries and has raised questions on the meaning of international water governance and its reach with regard to the allocation of shared responsibilities and management. In short, groundwater aquifers cross political and jurisdictional lines, heterogeneous, and sometimes conflicting, national and regulatory frameworks, so making collective management particularly challenging (see Figure 3.1: Cooley *et al.*, 2012).

The Guarani Aquifer is one such system (henceforth, GAS – Guarani Aquifer System) that supplies fresh water to Brazil, Argentina, Paraguay and Uruguay (Figure 3.1). As one of the largest aquifers in Latin America with an area of roughly 1,100,000 km², the GAS can store an estimated volume of 30,000 km³ of freshwater (Foster *et al.*, 2009). This aquifer is particular for several reasons. First, there is no evidence of conflict or dispute between either nations or sub-national jurisdictions (such as provinces and states) with regard to the aquifer and its uses. Second, because of its large size, the aquifer underlies many jurisdictional borders, so special emphasis must be placed on management structures at these border areas where conflict over use and use-rights can conflict. However, third, because of its massive scale, management of the aquifer presents particular challenges for states at temporal and spatial scales. The water quality is generally good, even around the densest human settlement (see Ribeirão Preto case study below). Problems when they arise tend to be highly localized, so failing to mobilize TBGWB (transboundary ground water basin) states collectively. The resource also appears to be inexhaustible, except in the northeast and southwest portions of the GAS. In light of climate change

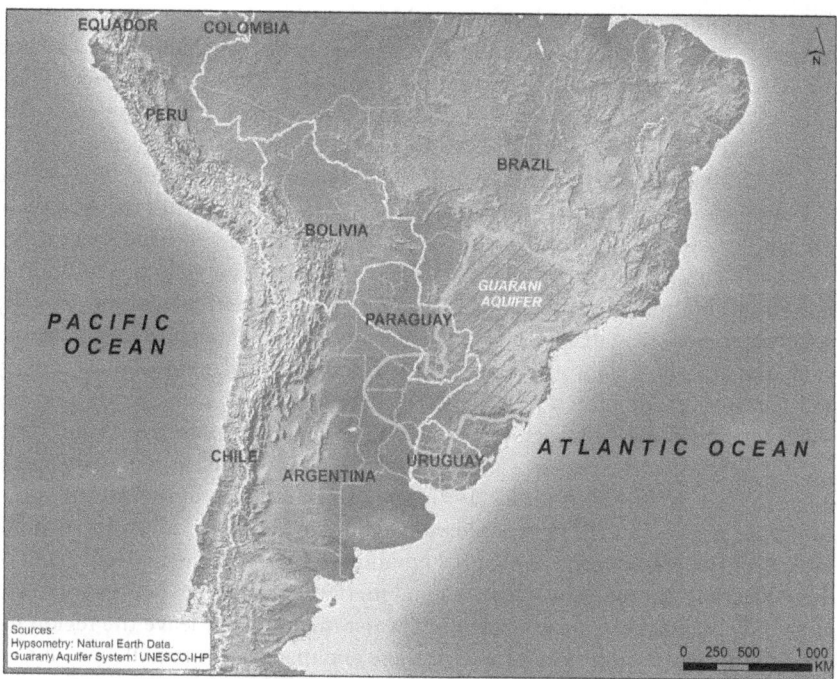

Figure 3.1 The Guarani Aquifer Basin.

projections of increased rainfall across the TBGWB, this is likely to further inhibit a sense of urgency in collective planning, use and management.

On 2 August 2010, the governments of Brazil, Argentina, Paraguay and Uruguay signed the Guarani Aquifer Agreement (GAA). According to Villar (2016b: 14), this agreement 'reaffirmed the applicability of international water law principles to aquifers and was the first agreement for transboundary groundwater developed under the influence of the UNGA Resolution 63/124 (2008)'. This resolution established the Law of Transboundary Aquifers. A combination of factors led to the creation of the GAA. Within the GAS, increasing demand for freshwater; increasing environmental costs associated with rapid urbanization, expanding commercial agriculture and mining; and anticipated adverse effects and significant challenges of climate change were dominant drivers. So, scientific reports about and citizens' experiences with recurrent droughts, point-source pollution and land degradation seem to have successfully mobilized states in the first instance. Beyond the region, the water world had been moving in the direction of transboundary water governance for some time – a natural follow-on from the growing emphasis on integrated water resources management (IWRM), good water governance and so on (Swatuk, 2002). Significant international involvement within the TBGWB states helped translate these global norms into projects, programmes and policies at national, sub-national and transnational levels.

Though signed in 2010 by all four nations, the GAA is not yet operational. To the contrary, since its inception, there is evidence of poor management headed by a lack of communication from all stakeholders including national and sub-national governments regarding the overall treatment of the aquifer. This chapter reflects on the challenges and opportunities presented by the collective management of a largely invisible resource, especially one so physically large as the GAS. Background on the laws of transboundary water which led to the Guarani Aquifer Agreement are initially discussed. Following this, the efficacy of the various stakeholder engagements to date are described. With a look at the city of Ribeirão Preto in Brazil, the chapter will show not only the local-level side effects, but reflect on the possible boomerang effects emerging from generally poor water resources management, particularly at the municipal level. In our view, the transboundary water agreement between Brazil, Argentina, Uruguay and Paraguay exemplifies a progressive and peaceful step on the road to proper management of the GAS. However, in the absence of conflict or a clear and present danger to the resource at a grand scale, it is not clear what will move states towards what Karen Litfin (1997: 169) labels a 'sovereignty bargain', i.e. a condition where the GAS states cede some of their individual sovereignty in the interest of the greater good of regional groundwater governance for the benefit of the collective. In our view, then, a combination of sovereign self-interest and dispersed challenges to local-level groundwater management leave the region ill-prepared for more serious resource access, use and management challenges ahead.

Evolution of a transboundary water agreement: the Guarani Aquifer Agreement

Transboundary water law has evolved as governments and policymakers became more knowledgeable about the science of water and the interrelationship between surface and groundwater (Eckstein and Eckstein, 2005). The International Law Association in 1966 adopted the Helsinki Rules on the Uses of the Waters of International Rivers, otherwise known as the Helsinki Rules. They represented the first comprehensive, international guidelines to regulate the use of transboundary rivers and their connected groundwater aquifers (Cooley et al., 2012). These principles highlight the importance of equitable and reasonable utilization, the dangers of pollution and a set of procedures for the prevention and settlement of potential disputes (ILA, 1966). These rules were followed by the Convention on the Law of the Non-Navigational Uses of International Watercourses (UN Convention) in May of 1997. This UN Convention is the strongest legal instrument regarding transboundary water management to date, becoming operational in August 2014 (Swatuk, 2017; Cooley and Gleick, 2011). These legal frameworks provided the necessary foundation upon which the GAA Nations built their agreement (Villar, 2016b).

The events that led to the creation of the Agreement provide some insight on the lack of implementation with regard to the development of a collective management framework from the GAA nations. From 2003 to 2009, the World

Bank, the Global Environment Facility, the IAEA, Germany's Federal Institute for Geosciences and Natural Resources, the Bank of Netherlands Water Partnership programme along with the four GAS nations developed the Project for the Protection and Sustainable Development of the Guarani Aquifer (Sugg *et al.*, 2015; OAS, 2009). This project was an ambitious initiative in South America for groundwater (Villar and Riberio, 2013). The six-year project (2003–2009) increased awareness of the GAS's characteristics and stimulated debate on groundwater management within the four countries at national, provincial and community levels (Villar and Riberio, 2013). At the same time, the Mercosur Common Market Council, otherwise known as the High-Level Group, was tasked to analyze the results produced from the Project and draft an agreement in the management of the GAS (Sindico, 2011). However, nothing substantive was formed from these discussions (Villar and Riberio, 2013; Sindico, 2011). On the international front, the UN General Assembly adopted the Resolution on the Law of Transboundary Aquifers in August 2008. The resolution mirrors the principles of the UN Convention, as it highlights the necessity of appropriate bilateral or regional arrangements for the proper management of their transboundary aquifers (UNESCO, nd). It is this framework that led to the Guarani Aquifer Agreement. The Agreement reiterates the values of state sovereignty, cooperation and flexibility, as well as a framework for dispute mitigation and settlement (Sindico, 2011). Today, the framework remains in place in order to mitigate future conflict.

Many authors have discussed the Guarani Aquifer with regard to the effectiveness of international water law and the role of water governance in this setting (Villar, 2016b; Cassuto and Sampaio, 2013; De Castro, 2012; Sindico, 2011). In a transboundary water-basin context, history has shown that countries that share resources possess more incentive to cooperate than to compete feverishly in order to maximize the use of the water (De Castro, 2012). The Guarani Aquifer Agreement is a regulatory framework and the first of its kind on this scale to follow this ideology. It is in effect the first of many steps needed to ensure the quality and sustainability of the aquifer. With a reliance on other international agreements, the GAA shows, on one hand, the active engagement of nations to refer to international laws and treaties. Furthermore, the GAA legitimizes the need for high-level negotiations and flexible facilitation which nations that rely on a transboundary watercourse can use as a point of reference in development of domestic laws that follow this scheme.

Although the GAA is almost a direct reiteration of UN Laws on Transboundary Aquifers and the UN Convention, Article 16 of the GAA, which discusses the institutional framework and dispute settlement mechanism, is worthy of a deeper review. The GAA does not lay out key principles that the four countries sharing the aquifer must take into account in order to foster sustainable management but does provide the institutional setting for the latter (Sindico, 2011). Prior to the GAA, the Plata River Basin Treaty was meant to provide the institutional structure and management of the aquifer. Article 15 of the GAA establishes this framework:

It is established under the Treaty of the Plata River Basin, and in accordance with the Article VI of such Treaty, a Commission comprised by the four Parties, which shall coordinate the cooperation among such Parties for complying with the principles and objectives of this Agreement. The Commission shall elaborate its own regulations.

> (GAA (n.d.) accessed at http://extwprlegs1.fao.org/docs/pdf/
> mul-143888English.pdf 23 March 2018; also, Sindico, 2011)

This article is paramount, as it focuses the responsibility for the management of the aquifer on GAA nations. Should this Commission be implemented, it would lead to the desired outcome of an integrated and collective water-management structure for the GAS. As transboundary water basins exhibit complexities in terms of scale and competing values on the uses of freshwater, Article 16 sheds lights on the importance of cooperation in water management and governance:

> The Parties shall settle disputes concerning the interpretation or application of the present Agreement in which they are part, through direct negotiations, and shall inform the body referred to in the previous Article over such negotiations.
>
> (GAA (n.d.) available at http://extwprlegs1.fao.org/docs/pdf/
> mul-143888English.pdf)

It is difficult to state the sole driving force that led to the consensus of the Agreement. Some authors state that the removal of Article 19, which suggests two options for dispute resolution, including unilateral and bilateral arbitration, proved too contentious an issue for the GAA nations (Cassuto and Sampaio, 2013; Sindico, 2011). Moreover, the failures of the Mercosur negotiations demonstrated the need for a higher regulatory body, such as the UN and the International Law Association, to facilitate productive negotiations in the discussions surrounding participation and collective management of the aquifer. On this stage, the GAA nations have begun the necessary steps to move towards an integrated water resource management (IWRM) approach. Let us look more closely at the roles that each nation plays in implementing these frameworks.

Major stakeholders of the Guarani Aquifer system

According to the GAA, each of the four nations have sovereign control over their portion of the Guarani Aquifer and its management will thus be governed by the states' individual laws and legislation. The GAA mandates that it is the responsibility of nations to manage and protect natural resources while, in accordance with the 1992 Rio Declaration on the Environment and Development, maintaining sustainable development for future generations (GAA, 2010). A review of domestic environmental legislation from each of these nations provides a deeper understanding of the extent of future cooperation and the Aquifer as a whole. On the national scale, the countries of the GAS have made many improvements in environmental policies, but many of these policies, though translated into law, have not been implemented (Villar and Ribeiro, 2011). As the evidence will show, the national environmental and water management policies of the four GAA nations are varied and issues related to boundaries even exist within nations themselves.

In Brazil, the local ideology is that the environment is a collective asset that is to be defended and protected for current and future generations (Garcia and Feldmann, 2016). The environmental legislation and enforcement in Brazil is shared between all levels of government: state, federal and municipal. Municipal environmental agencies are required to manage local activities that will have an impact within the respective municipality (Garcia and Feldmann, 2016). The Brazilian Institute of the Environment and Renewable Natural Resources is the federal agency that has jurisdiction over sensitive licensing issues such as nuclear power (Garcia and Feldmann, 2016). Finally, environmental agencies controlled by the state manage everything not encompassed by municipal or federal agencies. As a result the environmental agencies of Brazil are very fragmented. Although this means that many environmental licenses and management are localized at the municipal level, issues of watershed management and large aquifers will not be managed at the appropriate level that takes the entire system into consideration.

Argentina's Constitution provides the basis for environmental regulation in the country. Section 41 affords the right to all citizens to a healthy environment as well as the duty to restore any damage caused to the environment in the form of 'polluter pays' (Macchiavello and Sesto, 2016). The responsibility to enact environmental codes and legislation is left with the federal and provincial governing bodies (Macchiavello and Sesto, 2016). The General Environmental Act passed in 2002 represents the federal baseline for environmental policy (Macchiavello and Sesto, 2016). Thus provincial legislation can build upon this Act for increased environmental laws and provisions for sustainable development. The environmental policy in Argentina, although less fragmented than Brazil, due to the baseline established by the General Environmental Act, may be problematic when managing watersheds that are transboundary across provinces.

The Uruguayan Constitution plays a vital role in the environment policy of the country. Section 47 of the Constitution states that the environment should be protected from degradation, pollution and destruction, and should be restored when damage has come to it (Aldaz and Saracho, 2016). This assertion is reinforced by two more national bodies, the Ministry of Housing, Territorial Planning and Environment and municipal authorities when appropriate (Aldaz and Saracho, 2016). According to Uruguayan law there is a possibility of fines being levied against persons or entities liable for soil and groundwater contamination (Aldaz and Saracho, 2016). Therefore, in Uruguay the environment is legally protected at the federal and municipal levels. As previously mentioned it can be concluded that Argentina and Uruguay's Constitutions align with regard to the rights of their citizens to have a clean and healthy environment.

Outlining a brief overview of Paraguay's environmental legislation is difficult, as its legislation is limited with regard to environmental protection and water. The 1992 National Constitution uses vague and somewhat contradicting language when referencing the environment, articulating 'priority objectives of social interest: the preservation, the renovation and the improvement of the environment, as well as the conciliation with the human being' (International Business Publications, 2015). Although all citizens are afforded the right to live in a healthy environment, the use of words such as renovation and improvement leave room for interpretation that may lead to further degradation. In 2004 a Subterranean Water Law was passed but its presence in the literature is scant and only vaguely mentioned. Thus the effectiveness of this law is seemingly minimal at best and nonexistent at worst. Villar (2016b: 13–14) summarizes the legal basis for transboundary water management as follows:

> At the national level, the 2000s have brought with them the drafting of laws and plans for groundwater. For example, Argentina formulated the National Plan for Groundwater (2007); Brazil created several legal acts to include groundwater in water management and launched specific programs such as the National Groundwater Agenda19 and the National Program for Groundwater20; Paraguay established a specific law for water resources (Law nº 3.239/2007); Uruguay established a National Water Policy (Law nº 18.610/2009) to complement the Water Code (Decree nº 14.859/1978), created the Guarani Aquifer Management Plan (Decrees nº214/2000, 11/2001 and 295/2001) and established the Commission for the Guarani Aquifer (Decree nº 183/2013) as an advisory organ of the Water Resources Regional Council for the Uruguay River.

Overall, these nations have a strong basis for environmental policy, including water management, but have varying degrees of implementation and enforcement. In order to understand how the stakeholders interacted leading up to the GAA, all National Stakeholders are listed for the Project for the Environmental Protection and Sustainable Development of the Guarani Aquifer System mapped out in Table 3.1 (2009). The first participant in each column represents the most prominent stakeholder from each of the four countries. The number of participants from each county shows the impact and possibilities available to each nation. The potential impact and roadblocks to the management of the GAS are highlighted by the example of the largest stakeholder, Brazil, in the following section.

Stakeholder highlight on Brazil

As highlighted above, the Guarani Aquifer represents an important and strategic source of freshwater. Brazil, due to both its political and economic prowess in the region as well as the fact that more than 60 per cent of the Guarani Aquifer is within its political boundaries (OAS, 2009: 17), has taken

Table 3.1 National stakeholders of the Guarani Aquifer System

Brazil	Argentina	Uruguay	Paraguay
Secretariat of Water Resources and Urban Environment of the Ministry of Environment	Under-Secretariat for Water Resources of the Federal Ministry of Planning, Public Investment and Services	National Directorate for Water and Sanitation of the Ministry of Housing, Territorial Planning and Environment	Directorate General for Protection and Conservation of Water, Resources of the Secretariat of Environment
National Water Agency	Secretariat of Environment and Sustainable Development of the Nation	National Directorate for Environment of the Ministry of Housing, Territorial Planning and Environment	Technical Secretariat for Planning
Ministry of Foreign Affairs	Ministry of Foreign Affairs, International Trade and Worship	Ministry of Foreign Affairs	Ministry of Foreign Affairs
Secretariat of Environment of the State of Mato Grosso	Directorate for Living Resources of the Ministry of Ecology, Renewable Natural Resources and Tourism of Misiones	Municipal Intendancy of Salto	University Foundation for Agrarian Sciences of Itapúa
Secretariat of Environment, Cities, Planning, Science and Technology – Environment Institute of the State of Mato Grosso do Sul	Water and Environment Institute of Corrientes	Municipal Intendancy of Rivera	
Secretariat of Environment and Water Resources of the State of Goiás	Directorate for Hydraulics of the Municipality of Entre Ríos		
Water Management Institute of the State of Minas Gerais	Provincial Potable Water and Sanitation Service		
Secretariat of Environment	Provincial Water Administration		

continued

Table 3.1 Continued

Brazil	Argentina	Uruguay	Paraguay
Superintendency of Water Resources Development and Sanitation	Ministry of Water, Public Services and Environment of Santa Fe		
Directorate for Water Resources of the Secretariat of Sustainable Economic Development of the State of Santa Catarina	Municipality of Concordia		
Department of Water Resources of the Secretariat of Environment of the State of Rio Grande do Sul			
Department of Water and Electric Power			
Department of Water and Sewage			

Source: Derived from OAS (2009)

the lead on the GAA and management of the Aquifer (Wendland, Rabelo and Reohrig, n.d.). Brazil uses 87 per cent of the Aquifer's withdrawals, reflecting its high rate of urbanization and access to the aquifer (Wendland, Rabelo and Reohrig, n.d.; Cassuto and Sampaio, 2013). With the majority of the management falling to Brazil due to the aquifer's location, it is important to consider the country's role in the GAA and its national water-management processes, as they stand to have the most effect on the GAS (Cassuto and Sampaio, 2013). The ecological realities of the GAS must be aligned with a harmonized set of laws in Brazil, Argentina, Uruguay and Paraguay (Villar, 2015; 2016a). Brazil has a proportionally larger responsibility to implement and uphold appropriate water management strategies due to its potential to affect the quality of the GAS.

Within its legal system and government structure Brazil has multiple authorities split up between jurisdictions. For example, land-use zoning may be an indirect but important factor in groundwater usage and potential contamination. Some observers feel that the fact that municipalities in Brazil have autonomous power to decide on land-use zoning affecting contamination, recharge and withdrawal of the GAS causes inconsistencies in water management in the country (Cassuto and Sampaio, 2013). Similarly, Victor *et al.* (2014: 10) state,

Problems of overlapping jurisdiction extend beyond ownership rights, reaching into many matters that relate to water either directly or indirectly. For instance, issues in the areas of conservation of nature, soil and natural resources, environmental protection, and pollution control all have concurrent jurisdiction over the treatment of water.

Other observers, however, feel that multiple jurisdictions – national, state, municipal, inter-ministerial and departmental – are not a hindrance to coordinated groundwater planning, use and management. For example, Patole (2015) highlights the steps taken to ensure sustainable use of groundwater in Brazil. The main steps include the following. First, the 1988 Federal Constitution ended private ownership of water and made it a public good under control of the central state. The Union government sets minimum legal standards for all states to follow. As set out in the Constitution, the Union has jurisdiction over transboundary (inter-state within Brazil; between countries) surface waters, while the states have jurisdiction over intra-state surface water and groundwater. At the same time, however, all four countries have accepted in principle the terms of UN Res 63/124 of 2008 on transboundary groundwater law. Most important here are the terms 'equitable and reasonable use' as well as protection of the resource against pollution.

Second, in 1997, Brazil passed Law No. 9433 – Brazilian National Water Resources Policy in which the groundwork was laid for an IWRM approach in line with emerging global trends (Swatuk, 2002). Third, 10 years later, in 2006, Brazil launched its National Water Resources Plan (PNRH is the Portuguese acronym) where groundwater is treated in its own chapter. Fourth, Patole (2015: 13–14) highlights several Resolutions passed by the National Water Resources Council (NWRC) pertaining to the use and management of groundwater: Resolutions 15 (general guidelines), 16 (licensing), 17 (integration), 22 (multiple use) and 92 (harmonization of laws and practices). The author also goes on to describe similar actions taken by other Federal and State water resource-focused institutions focused on related important issues such as pollution control. Taken together, Patole argues, Brazil has closed the gap between surface and groundwater planning and management.

Given groundwater's importance to all facets of Brazil's economy, the national stakeholders are many, with urban centres, rural agriculture and mining being the main ones. Across the GAS, there are an estimated 500 cities, most of these in Brazil, where the bulk of economic activity takes place. Eighty-seven per cent of the GAS population resides in Brazil, constituting 43 per cent of Brazil's total population – roughly 90 million of an estimated 186 million people (OAS, 2009). Economic activity in Brazil accounts for approximately 80 per cent of the GDP across the GAS (Argentina accounts for 18 per cent; Uruguay for roughly 1.5 per cent; Paraguay for less than 1 per cent) (OAS, 2009: 96). While secondary (33 per cent, e.g. manufacturing) and tertiary (53 per cent, e.g. services) industry accounts for the bulk of Brazilian economic activity in the GAS, primary industrial activity (e.g. agriculture and mining, 13.6 per cent) plays a significant role as well (OAS, 2009: 96).

Villar (2016a) reports that due to the increasing degradation of surface waters due to pollution, and to recurrent droughts across the country, there is something of a well-drilling bonanza underway at all scales of Brazilian society. She reports that 40 per cent of municipalities depend solely on groundwater for their supply. A significant proportion of medium and large cities, as well as rural households, are increasingly dependent upon groundwater, so totalling roughly 87 million people (Villar, 2016a: 93). She also reports that despite laws requiring registration, it is estimated that no more than half of existing wells have been registered (Villar, 2016a: 91).

Climate change is hypothesized to increase the amount of rainfall received across the GAS (OAS, 2009). The GAS lies roughly between 56°W and 48°W longitude, with average rainfall varying at 24°S latitude: >1,300 mm/a north of 24°S and >1,600 mm/a south of 24°S (OAS, 2009). These are significant amounts of rain that fall intensely and seasonally. Hardened soils from expanding urban settlements and rural land degradation will increase runoff and decrease infiltration, so exacerbating both surface and ground water quality and quantity issues (Villar, 2016a). Yet, as suggested by Villar (2016a), the actual management of these resources, especially groundwater, is lacking.

A number of lawmakers have proposed to amend the Constitution in order to clarify the Federal Government's role in governing groundwater, but pushback was seen from agribusiness associated interests (Cassuto and Sampaio, 2013). By maintaining decentralized management of groundwater resources, agribusiness hopes that states are more receptive to their land-use zoning and other benefits (Cassuto and Sampaio, 2013). At the same time, environmentalists highlight the continuing degradation and deforestation of landscapes throughout the GAS. For example, the percentage of land within the GAS that is classified as thick or degraded forest has declined from 42 per cent in the 1973–1980 period to less than 8 per cent today. This is due primarily to agricultural expansion (OAS, 2009). The implications for aquifers and rivers are clear: increased runoff as well as increased build-up of pesticides and fertilizers in the soil and in water bodies will place increased stress on natural systems, so impacting human-made systems of access, use and management.

Although conflicting interests in the management of groundwater in Brazil is evident, states within the GAS have enacted controls to attempt to manage access and usage of the aquifer (Cassuto and Sampaio, 2013). State legislation can only go so far and in the case of a transboundary aquifer such as the Guarani, management should be done at the national scale in order to better coordinate with the three other nations with vested interests in the GAS (Cassuto and Sampaio, 2013). Cassuto and Sampaio (2013) assert that although an emphasis is placed on Federal management of transboundary groundwater, such as the Guarani Aquifer, for the sake of diplomacy and international relations, state and municipal actors should not be left out, as their local interests are important in the management of water resources. As shown by Patole (2015), the appropriate architecture seems to be in place; however, as suggested by Villar (2016a) and others, implementation in support of IWRM-oriented practice is lacking.

Case study: Ribeirão Preto

As is evident throughout the water world, excellent laws and policies do not necessarily translate into effective practice. Brazil is no exception in this case. Persistent, poorly regulated and unpunished misuse of surface and groundwater resources means that many cities, villages and rural areas are suffering local-level side effects such as lack of access to fresh water. This abiding contradiction between policy and practice may, in our view, lead to a boomerang effect particularly at the level of local municipalities (Sindico, 2011). Let us turn briefly to the example of the municipality of Ribeirão Preto in Brazil, which is suffering from the impacts of poor management of the aquifer.

Ribeirão Preto is a city of some 665,000 people, though some authors suggest the total is closer to one million. It is located at 21.1704°S latitude and 47.8103°W longitude, in the southeast of São Paulo state. The city has a long history steeped in the region's development as an important agricultural producer, today including sugar cane, coffee and oranges (Foster et al., 2009). The city itself is home to a wide industrial base and has lately become a centre for technological innovation (Sindico, 2011). Its urban footprint has expanded significantly over time, with a concomitant hardening of the urban landscape and degradation of outlying areas. Such have been the challenges of sustainable water supply that the city has become part of a World Bank pilot project in water resources management in the GAS (Foster et al., 2009). As of 2010, the city was drawing 13,400 m^3/h from 102 wells (Sanches et al., 2010: 1). According to Foster et al. (2009),

> [g]roundwater development and water-table lowering have reduced, and largely eliminated, natural groundwater discharge to streamflow (and replaced it by wastewater discharge).... Contemporary groundwater recharge is exceeded by abstraction since, over a large area across the city, groundwater levels have fallen by an estimated 30 m–40 m since 1970.

Indeed, the authors point out that by 2009 groundwater levels were so low that stream effluent had reverted to influent, so exposing the aquifer to pollution. As the demand grows and the aquifer is unable to recharge, Ribeirão Preto runs the risk of being out of water by 2050 (Foster et al., 2009). In 2013, it was reported that the city had overdrawn its groundwater to such an extent that it had to pursue a surface-water transfer scheme. The US$300 million project would tap the waters of the Pardo River. Foster et al. (2009: 18) highlight the challenges to sustainable water supply in Ribeirão Preto as follows: appropriate land-use planning; clear understanding of the impacts of 'urban sanitation measures, industrial activities and agricultural practices' on water resources; better practices related to the 'urban water cycle' including wastewater reuse; demand management as it relates to groundwater abstraction (currently at 350 l/cap/day projected to rise 'by a further 187 mm^3/a in 2030'); and exploration of different abstraction sites less susceptible to pollution.

Limited progress had been made in a number of ways according to Foster *et al.* (2009), including the promulgation of São Paulo State Law and Ribeirão Preto Municipal Decrees related to groundwater management, and joint groundwater monitoring efforts with various partners. Organizations such as Comitê da Bacia Hidrográfica do Pardo are attempting to reduce water demand within the region through environmental education (Sindico, 2011). The Instituto Geologico de São Paulo has partnered with international non-governmental organizations to analyze the more technical aspects to better understand the severity of the situation from pollution to vulnerability (Sindico, 2011). Clearly, these efforts are better than nothing, but in our view are the very least authorities could do.

The Ribeirão Preto case is instructive in several ways. It is indicative of the patterns of growth and resource use across much of Brazil. To this point in time, groundwater has been considered an inexhaustible resource that is of good quality and so relatively inexpensive to supply. As shown above, many cities across Brazil are now opting to tap groundwater because surface waters are so polluted. Ribeirão Preto, interestingly, is moving in the opposite direction. This is a warning to cities and towns across the GAS: overuse your groundwater at your own peril. The case study therefore provides a cautionary tale about the limits to growth, and how surface water, groundwater, land use and urban/industrial development constitute a complex web of resource vulnerability. A third value of the case study is to show how little integrated the landscape is in terms not only of resource use but in terms of resource planning, policy, law and project implementation. Disintegrated and unchecked growth across the GAS is tantamount to 'death by a thousand cuts'. While you may not feel the individual incision, their cumulative impact adds up. Put differently, in the absence of an operationalized regional GAA, recurrent crises will be treated separately at the site. Over time, they will add up, causing multiple forms of stress across the GAS. Must sovereign states wait until such time that only region-wide crisis forces their cooperative hand? This returns us to a consideration of the value of the GAA and an IWRM perspective.

Next steps and considerations

Strong institutions build the firm foundation for lasting peace in areas of transboundary water. The GAA is an important mandate in establishing a lasting institutional framework in the area, but additional constructive strategies are needed to establish appropriate regulatory and enforcement bodies. A coordinated effort between Brazil, Argentina, Uruguay and Paraguay to enact national agencies that can manage the GAS is pertinent to securing the future of the system for many generations. Each of the four countries have seen advancement in recent years for environmental legislation protecting specific aspects of their nation's natural resources for use by their populations. Despite the region being inundated with issues such as rising sea levels due to climate change, point-source pollution and the growing fear of drinking water scarcity, the countries overlying the aquifer have remained peaceful. In the last 50 years only 37 disputes involving transboundary water have caused violence while 157 have resulted in peaceful treaties (Wolf *et al.*, 2003). Conflict related to resource scarcities would seem inevitable and normally lead to various forms of political instability, while pressuring governments to plan for action. For Wolf *et al.*, (2005: 85), '[W]ater is an irritant and a unifier'.

As is well known, states fight and cooperate over the same things: quantity, quality and the means of addressing these concerns (Swatuk, 2017). The Guarani Aquifer case is unique in that a transboundary groundwater agreement was arrived at in the absence of any inter-state conflict. As Villar (2016a; 2016b) points out, however, there has been a management malaise in the years following the agreement. One might convincingly argue that the early great strides were made when human and financial resources beyond the region were readily available. It is doubtful that the GAS states would have undertaken the Guarani Aquifer System Project (2003–2009) with their own resources (OAS, 2009). In Villar's view (2015; 2016b), the Guarani Aquifer management process has followed three phases: the first phase involving 'epistemic community studies'; the second phase involving international organisations and regional states; and a third phase involving the GAS states themselves. Typical of many transboundary arrangements elsewhere in the world (see Swatuk, 2017 for the Southern African case; see Mirumachi, 2015 for examples from the Middle East, South Asia and Southern Africa), the Guarani Aquifer arrangement has fallen at the final hurdle. When left to states, sovereignty is a fundamental deterrent to deepened cooperation. So, the GAS states are happy to conduct joint studies, but reluctant to ratify the GAA into national law, and unwilling to establish a Transboundary Groundwater Basin Commission. Brazil, in particular, as the basin hegemon, sees little need to defer to the interests of the three smaller actors, one of whose parliament has rejected the agreement altogether.

With climate change impacts looming on the horizon, and with unchecked development occurring all across the GAS, one would hope that the GAS states would revisit the GAA, establish a Commission, agree on a dispute settlement mechanism and work towards I(G)WRM – integrated groundwater resource management. Instead, it appears that they are content to muddle through, leaving issues of supply and sustainability to be dealt with at the points of extraction. Should conflict arise, it is likely to be local and, for the national state, with the application of technology and money – e.g. the water supply system in Ribeirão Preto – likely to be contained.

References

Aldaz, A., and Saracho, M (2016). Uruguay: Environmental and climate change law 2016. *International Comparative Legal Guides*. Available at: www.iclg.co.uk/practice-areas/environment-and-climate-change-law/environment-and-climate-change-law-2016/uruguay. Accessed 26 March 2018.

Cassuto, D. N., and Sampaio, R. S. (2013). Hard, soft and uncertain: The Guarani Aquifer and the challenges of transboundary groundwater. *Colorado Journal of Environment Law and Policy*, 24(1), 1–41.

Castro, D. D. (2012). The shared management of the Guarani Aquifer: The South American example in global governance over water resources. *Yearbook of International Environmental Law*, 22(1), 140–157.

Cooley, H., and Gleick, P. H. (2011). Climate-proofing transboundary water agreements. *Hydrological Sciences Journal* 56(4): 711–718.

Cooley, H., Christian-Smith, J., Gleick, P. H., Allen, L., and Cohen, M. J. (2012). Climate Change and Transboundary Waters. In: P. H. Gleick (ed.), *The World's Water, Vol. 7*, Oakland, CA: The Pacific Institute, 1–22.

Eckstein, Y., and Eckstein, G. E. (2005). Transboundary aquifers: Conceptual models for development of international law. *Ground Water*, 43(5), 679–690.

Foster, S., Hirata, R., Vidal, A., Schmidt, G., and Garduno, H. (2009). The Guarani Aquifer initiative – towards realistic groundwater management in a transboundary context. The World Bank: Water Partnership Program.

Garcia, L. P., and Feldmann, R. (2016). Brazil: Environment and climate change law 2016. *International Comparative Legal Guides*. Available at: www.iclg.co.uk/practice-areas/environment-and-climate-change-law/environment-and-climate-change-law-2016/brazil. Accessed 26 February 2016.

Guarani Aquifer Agreement (GAA) (2010). Available at: http://extwprlegs1.fao.org/docs/pdf/mul-143888English.pdf. Accessed 26 March 2018.

International Business Publications (2015). *Paraguay: Ecology and Nature Protection Handbook, Vol. 1, Strategic Information, Programs and Regulations*. Washington, DC: IBP for Government of Paraguay.

International Law Association (ILA) (1966). Helsinki rules on the uses of the waters of international rivers. Available at: http://webworld.unesco.org/water/wwap/pccp/cd/pdf/educational_tools/course_modules/reference_documents/internationalregionconventions/helsinkirules.pdf . Accessed 26 February 2016.

Liftin, K. (1997). Sovereignty in world ecopolitics. *Mershon International Studies Review*, 41, 167–204.

Macchiavello, G., and Sesto, L. (2016). Argentina: Environment and climate change law 2016. *International Comparative Legal Guides*. Available at: www.iclg.co.uk/practice-areas/environment-and-climate-change-law/environment-and-climate-change-law-2016/argentina. Accessed 26 March 2018.

Organization of American States (OAS) (2009). *Guarani Aquifer: Strategic Action Program. Acuifero Guarani: programa estatégico de acción – Bilingual edition – Brazil, Argentina, Paraguay, Uruguay*. Bilingual edition. Brazil: OAS.

Patole, M. (2015). Brazilian groundwater law – abstraction and pollution controls. (unpublished) DOI: 10.13140/RG.2.1.2358.5365.

Sanches, S. M., Vieira, E. M., do Prado, E. L., and Takayanagui, A. M. M. (2010). Qualidade da água de abastecimento público de Ribeirão Preto em área de abrangência do Aquifero Guarani: determinação de metais e nitrato. *Ambiente e Agua*, 5(2), 1–15.

Sindico, F. (2011). The Guarani Aquifer System and the international law of transboundary aquifers. *International Community Law Review*, 13(3), 255–272.

Swatuk, L. A. (2002). The new water architecture in Southern Africa: reflections on current trends in the light of 'Rio + 10'. *International Affairs*, 78(3), 507–530.

Swatuk, L. A. (2017). *Water in Southern Africa*. Pietermaritzburg: UKZN Press.

Sugg, Z. P., Varady, R. G., Gerlak, A. K., and Grenade, R. D. (2015). Transboundary groundwater governance in the Guarani Aquifer System: Reflections from a survey of global and regional experts. *Water International*, 40(3), 377–400.

Victor, D. G., Almeida, P., and Wong, L. (2015). Water management policy in Brazil. ILAR Working Paper No. 21 (March). San Diego, CA: School of International Relations and Pacific Studies.

Villar, P. C. (2016a). Groundwater and the right to water in a context of crisis. *Ambiente and Sociedade*, 19, 85–102.

Villar, P. C. (2016b). International cooperation on transboundary aquifers in South America and the Guarani Aquifer case. *Revista Brasileira de Política Internacional*, 59(1), 1–20.

Villar, P. C. (2015). *Aquíferos Transfronteiriços: Governança das Águas e o Aquífero Guarani*. Curitiba: Juruá.

Villar, P. C., and Ribeiro, W. C. (2013). The agreement on the Guarani Aquifer: cooperation without conflict. Available at: www.globalwaterforum.org/2013/09/02/the-agreement-on-the-guarani-aquifer-cooperation-without-conflict/. Accessed 12 March 2016.

Villar, P. C., and Ribeiro, W.C. (2011). The agreement on the Guarani Aquifer: a new paradigm for transboundary groundwater management? *Water International*, 36(5), 646–660.

Wendland, E., Rabelo, J., and Roehig, J. (n.d.). Guarani Aquifer System – the strategical water source In South America. Cologne: Institut für Tropentechnologie (ITT). Available at: www.geologiadelparaguay.com/Guaran%C3%AD-Aquifer-System.pdf. Accessed 26 March 2018.

Wolf, A. T., Kramer, A., Carius, A., and Dabelko, G. D. (2005). Managing water conflict and cooperation. In: *State of the World 2005: Redefining Global Security*. Washington, DC: Worldwatch Institute, Chapter 5

Wolf, A. T., Stahl, K., and Macomber, M. F. (2003). Conflict, cooperation, and university support for institutions in international river basins. Presentation at the annual meeting of the International Studies Association.

4   Shifting waters

Exploring how changing water discourse has ignored the impacts of the Farakka Barrage

Rija Rasul, Stephen Little, Zoya Khan and Larry Swatuk

Introduction

This chapter provides a cautionary tale regarding the often-discounted dangers, difficulties and challenges of dealing with water insecurity from the perspective of state sovereignty and through the application of capital and technology for big infrastructure (Zeitoun *et al.*, 2016). Much has been written about India in this context (Swain, 1996, 1993). This chapter focuses on the Farakka Barrage, and argues that a continued focus by the region's state-level actors on technological 'fixes' to natural fluctuations and climate change–affected variations in water availability will present South Asia's states with myriad local-level side effects (e.g. displaced peasants) and boomerang effects (e.g. local-level violent clashes between government and citizens; regional level political animosities that stand in the way of collective approaches in the service of benefit sharing). In addition, the national governments of India have, over time, tended to securitize water management (Mirumachi, 2015), meaning that proposed actions to be taken in the service of 'water security for the state' must be taken and are not to be disputed. To oppose a proposed action on any grounds, therefore, is to risk turning a possibly technical issue into an issue to be treated through high level inter-state channels such as foreign policy and national defence. Such persistence of high-modern infrastructure tied to high politics bodes ill relative to the flexibility and adaptability necessary for climate change– induced extreme variability.

Much like the Nile, the Amazon, and other historically significant rivers, the Ganges has been both a figurative and literal life-giving artery running through South Asia. The river has not only been a cultural pillar of the region, but also a primary economic driver, turning hydroelectric turbines and hydrating crops. In response to mounting pressures from downstream states in the 1950s, India moved forward with plans to construct the Farakka Barrage in 1961, signalling what some thought was the beginning of a more equitable transboundary water-sharing agreement between present-day Bangladesh (formerly East Pakistan) and India. However, this was premature. After years of longstanding disputes, the same arguments are imbued with new salience when quotas are transgressed or water stresses emerge (see Kawser and Samad, 2016 for details). As environmentalism consolidates itself as a mainstream framework to analyze ecological issues, poor management has been largely superseded by a focus on climate change stressors. This has resulted in negative local-level side effects – both social and environmental – as well as boomerang effects – in terms of strained political relations – in both nations. For India, diminished water resources are attributed to climate change, increasing the nation's reliance on Ganges water-training projects for crop irrigation. This has reduced river flow, forcing India to infringe on water agreements to maintain the Kolkata Port, one of the largest shipping ports in the world. Bangladesh, by contrast, has become focused on the interlinkages between water-related problems in western regions and climate change. However, this ignores the fact that Bangladesh has relatively low water stress, outside of the January–May dry season, and is still subject to riparian agreements and upstream regulation in India. While climate change should certainly not be ignored at any scale, the climate lens has not improved transboundary water sharing. In fact, it is preventing the root causes of water stress from being prioritized and seen for what they are: poor management and poorly conceived sharing agreements, factors that pre-date and exacerbate climate-induced stress. In chronological order, this chapter will illustrate how domestic, regional and international drivers are influencing the ways in which the climate change narrative takes shape and affects agreements and management practices around water resources.

Preconditions and construction (1950–1977)

Originating in the Himalayas, the Ganges River is approximately 2,500 km long and lies within the Ganges-Brahmaputra-Meghna (GBM) basin (Salman and Uprety, 2002). '[W]ith a total area of just over 1.7 million square kilometres, [it] is distributed between India (63%), China (18%), Nepal (9%), Bangladesh (7%) and Bhutan (3%)' (Hossain et al., 2015: 1). It is estimated that close to 630 million people reside in the basin – about one-third of Bangladesh's population and half of India's – many of whom depend on the Ganges for daily needs (Salman and Uprety, 2002). The entire system has a single terminus: Bangladesh. In addition to being integral to the survival of millions, the Ganges is sacred in the Hindu tradition. Not only are many towns that are considered holy stationed along the riverbanks, but the water itself is regularly used in religious ceremonies as well. It is believed that bathing in the river cleanses one's sins, and many float the bodies of their loved one's down as per traditional Hindu burial rites. However, this is heavily contributing to pollution and water contamination, risking the livelihoods of those who depend on the river (Conaway, 2015). Multiple cleaning efforts, including a multi-billion dollar initiative funded in part by the World Bank, have failed, as they have not delivered on goals (Das and Tamminga, 2012). In addition to the pollution problem, water sharing is a cause of immense tension between India and Bangladesh, and has been for decades. Allocation and distribution of water resources, especially during the five-month dry season, present one of the greatest challenges for sustainable management of the basin. Flooding and drought in the monsoon and dry seasons, respectively, occur naturally, and as a result, during the wet season water is plentiful, but both countries become water stressed during the dry season (Rahaman, 2006). Despite conflict-resolution attempts dating back over fifty years, India and Bangladesh have failed to reach an amicable solution. The rise of climate change and its adverse effects are undoubtedly creating new challenges and contributing to the problem as well. If the Gangotri Glacier (which feeds the Ganges) and other glaciers in the Western Himalayas begin to melt at unprecedented rates, it is expected that for periods of time, the Ganges could become ephemeral (Earle *et al.*, 2015). Approximately 70 per cent of Ganges water comes from the Gangotri Glacier during the dry season, and a shrinking glacier threatens millions (Earle *et al.*, 2015). Although climate change has recently presented new challenges, water distribution and management of the basin remain at the heart of the problem along the watercourse and are the underlying reasons for this ongoing conflict between India and Bangladesh.

In 1961, India officially announced its intention to build the barrage, to be located approximately 17 km from the India–Bangladesh border (Wolf and Newton, 2007). Rooted in the high-modern 'hydraulic mission' (Allen, 2003), the Barrage was completed in 1975 with the primary purpose of diverting 1,132 m³/sec of water out of an average dry-season flow of 1,416 m³/sec from the Ganges into the Hooghly River and down to Kolkata Port (Wolf and Newton, 2007). This original design was intended to serve as a 'low cost' means of flushing out the silt build up at the port. As the upper riparian state, India pursued this project in its own national interest. As a result, Bangladesh worried about the lost water supply for its own downstream needs. Despite numerous meetings between the two parties during the 14 years it took to construct the barrage, points of contention were never resolved. In 1972, the Prime Ministers of both states signed a declaration to agree to a mutually accepted resolution and to allow the Joint River Commission, a group established by both India and Bangladesh, to recommend solutions that met the needs of both parties (Salman and Uprety, 2002). This marked the genesis of the modern bilateral water agreements that exist today, conceptualizing water primarily through the lense of mutual water security (Mirumachi, 2015; Swain, 1996).

Post-construction (1977–1996)

Bilateral agreements

Although the 1974 resolution brought with it the promise of improved cooperation, upon completion in 1975 it was realized that flow rates were insufficient to meet both the needs of Bangladesh and of the Kolkata Port. Additional methods were needed to augment flow (Salman and Uprety, 2002). The two potential options were either through storage within the Ganges basin, which Bangladesh supported, or through diverting the water at Farakka, which was preferred by India (Wolf and Newton, 2007). Meetings continued, while India simultaneously continued to use the Barrage to divert water. In 1976 Bangladesh lodged its first formal complaint to the United Nations in protest of India's actions; the UN General Assembly responded by adopting a resolution calling upon both parties to quickly come to a fair and agreed upon solution (Rahaman, 2006).

A year later in 1977, the UN facilitated an agreement between India and Bangladesh, the Ganges Water Agreement. This five-year agreement covered water sharing at the Farakka Barrage, and potential long term solutions for augmentation of the dry season flows (Rahaman, 2006). According to Kawser and Samad (2016),

> [t]he share of India ranged from 40.7 per cent between January 1–10 to 37 per cent in the leanest period between April 21–30 and that of Bangladesh between 50.4 and 52.8 per cent respectively in the same period.

Bangladesh was guaranteed a minimum quantity of water in each 10 day period of lean season irrespective of the discharge available at Farakka. The agreement also established a Joint Committee, whose responsibilities included to measure, record, and regulate daily flows at Farakka (Rahaman, 2006). However, both parties failed to reach any agreement on dry-season augmentation within the defined time period, and when the agreement expired in 1982, it was not renewed.

After numerous efforts to salvage the agreement, India continued to unilaterally withdraw water from the Farakka, ignoring Bangladesh's calls to develop a novel and equitable solution. The response from India was unwelcoming. Eventually the issue was put to United Nations General Assembly and Commonwealth Heads of Government Meeting for mediation. Up until this time, the combination of India's perception of water as a bilateral issue and the unfavourable political relationship between the two countries stymied many possibilities for successful negotiation (Kawser and Samad, 2016). As such, the election of Bangladesh Awami League improved the political environment, which was a factor leading to the adoption of the next and last formal treaty that would be signed by both parties in 1996. The 1996 Ganges Water Treaty, a 30-year long agreement, was historic in the sense that it was considered a modern update to this decades-old conflict. Notable changes in this treaty included the establishment of a new method for the distribution of Ganges waters during the dry season, and this treaty additionally recognized Bangladesh's rights as a downstream riparian state (Wolf and Newton, 2007). However, despite international intervention and water laws encouraging cooperative methods, the primacy of power politics and economic dominance remained the key considerations for regional cooperation (Mirumachi, 2015).

Shifting global paradigms

The 1992 Rio Earth Summit, or the United Nations Conference on Environ-ment and Development (UNCED), changed the global narrative of develop-ment, recognizing the need for governments to take environmental impacts into account when making economic decisions. Businesses and governments were encouraged to make eco-efficiency a guiding principle, and the Summit's message was that attitudes and behaviours needed to change in order to decrease the stress on the environment (United Nations, 1997). It focused on four points:

- Patterns of production – particularly the production of toxic components, such as lead in gasoline, or poisonous waste – are being scrutinized in a sys-tematic manner by the UN and Governments alike;
- Alternative sources of energy are being sought to replace the use of fossil fuels, which are linked to global climate change;
- New reliance on public transportation systems is being emphasized in order to reduce vehicle emissions, congestion in cities and the health problems caused bye polluted air and smog;
- There is much greater awareness of a concern over the growing scarcity of water.

The Earth Summit resulted in an environment action plan for a century after 1992: Agenda 21. It formed the basis for a new international partnership for environmental protection and sustainable development, focusing on issues such as aid in the developing world and resource degradation (Meakin, 1992). Fol-lowing the Summit, narratives among UN members concerning water and dams diverged from the state-centric view to one that embraced human and environ-mental security as more significant pillars. To this point, Moudud (1998: 2) notes that

> the Earth summit in 1992 [convened] UNEP's Global 500 Laureates of the SAARC [South Asian Association for Regional Cooperation] … for cooperation … to educate our people and our policy makers to take the issue as a humanitarian and ecological issue.

This may have helped foster agreement between India and Bangladesh when crafting the 1996 Ganges Water Treaty.

Post-1996 agreement (1996–present)

Effects: social, political, ecological

The broadening of environmentalism in the 1990s has increased focus on the impacts of climate change as an exogenous factor that affects both state and human security. As yearly global temperatures surpass known records and ocean levels rise, climate change has attracted the attention of even those who were previously disinterested. While these changes certainly demand attention, this has had a negative effect, whereby policymakers and relevant stakeholders are focusing on state-specific interventions in response to climate change-induced scarcities while ignoring the real-time negative impacts of poor planning, resource development and management. In terms of issues stemming from poor management and the Barrage itself, the stalemate between the two countries left the Barrage's effects ignored. This has had different implications in each country. For Bangladesh, the Barrage has had a hand in increasing water salinity, siltation, flooding and loss of livelihood within western regions due to its effect on water flows and riverbank degradation (Hossen, 2015). Conversely, India has become more reliant on river training and linking projects to supplement water-impoverished regions, blamed in part on climate change. While this is intended to improve resiliency, it is making India's national water strategy more reliant upon the Barrage itself.

Bangladesh is rightly identified as one of the countries most vulnerable to climate change due in large part to its coastal position, low-lying altitude (Ministry of Environment and Forests, 2009), and the fact that 70 per cent of the country constitutes the floodplain for the GMB river basin system. At the same time, it is undeniable that the Barrage has also created hydrological vulnerabilities. The Chapra region, for example, has suffered congestion in the Gorai river due to increased silt deposition (Hossen, 2015). This has prevented adequate drainage, which caused flooding and ongoing displacement of this community. However, what is occurring here is not unique to this locality, as the number of displaced people in the greater southeast region has reached at least 10 million since the construction of the Barrage (Hossen, 2015). For the population remaining throughout the wet season, the transgression of water quotas by India has further affected the availability of water for crops (Hossen, 2015). And yet, instead of discussing the Barrage, the policy narratives examining the causal chains of these problems are heavily skewed towards climate change (Ministry of Environment and Forests, 2009). This is despite the fact that extreme events – drought and flood – are normal, and regularly overwhelm existing systems of management. According to research by Gain and Giupponi (2014), hydrological data shows sharp declines in minimum flows since the barrage was built, so worsening dry-season water scarcity.

The problem lies with improving nationwide access to water of sufficient quality and quantity for individuals and businesses. With Bangladesh's booming population growth and growing water demand from irrigation agriculture, water regulation must incorporate socio-economic matters, such as urbanization, population growth and farming practices. To accomplish this task, authorities need to not only understand water-related scarcity, crowding and flooding as anthropogenic problems, but also devise governance structures and policies to manage water at all levels of government. In this respect, the nation is a latecomer, as the country did not have any nationwide irrigation management policies until the 2013 National Water Act (UN Water, 2013; WWF, 2013). Given that the southwest/Gorai region of Bangladesh is largely supplied by water-flows from tributaries off the Padma River (as the Ganges is called in Bangladesh), it is critical to understand that, regionally speaking, flood and drought must be managed in accordance with water flows from the Barrage. Bangladesh recognizes this to an extent in their Rio + 20 National Report on Sustainable Development (MOEF GPRB, 2012: 52), saying:

> The Indo-Bangla Joint Rivers Commission deals with sharing and management of transboundary river waters, which is a very politically sensitive issue as it involves negotiations with India. Bangladesh shares 54 common rivers with India. Large scale withdrawal of water by India, the upper riparian, has been causing many of the trans-boundary rivers to carry very low flows or dry up with attendant land degradation, drought, ecological degradation, loss of habitat and agricultural decline in lower riparian Bangladesh.

In India, dramatic economic and population growth has increased water crowding and demand. As such, the state has searched for alternative sources of water, ways to improve efficiency and mitigate water loss. As part of this, climate change has been identified within India's National Water Mission (NWM) as a key contributor to diminished groundwater and surface water flows. The response has been The National River Linking Project (NRLP), a large-scale project that would link the major rivers throughout India, theoretically mitigating water scarcity by moving water from 'water-rich' to 'water-poor' basins. It would also redirect water from the transboundary rivers it shares with Bangladesh, pulling water from both the Ganges and Brahmaputra to other Indian regions. Not only will this greatly reduce Bangladesh's water security, it will alter the ecological systems of all impacted river basins (Bandyopadhyay and Perveen, 2003). Further to ecological impacts, the NRLP would potentially displace 500,000 people (Amarasinghe, 2009). The desire to chase large infrastructure projects, instead of increasing efficiency in existing projects or improve irrigation and agricultural practices, indicates that the Indian government is rarely identifying the root stressors to groundwater and river flow (Thomas *et al.*, 2014). The framing of the end goals for this project represents the persistently high level of denial that India has shown towards its impacts on water, the environment and the socio-economic welfare of its citizens and the hydrocracy's unflagging belief in technological solutions.

To understand these impacts, India only has to look to the Farakka Barrage and how its construction and ongoing operation has altered the natural river flow in Bangladesh and India. Erosion and silt deposition, known as *Char*, has caused Ganges waters to meander, causing displacement and internal migration in regions of close proximity to the barrage (Kumar-Rao, 2015). While studies have shown the migrating path of the river, there has been minimal acknowledgement of the issue or its cause by the Indian government (Thakur *et al.*, 2012). It has been postulated that due to the Ganges' capacity to erode the river banks, it may eventually bypass the Farakka, causing widespread damage and economic loss (IWMI, n.d.). However, methods to dampen the potential severity of the issues have focused on technocratic solutions to hydrological changes, not progressive management strategies.

Towards sustainability

Recent global events, reflecting current ideologies, have also impacted the narrative around water security in India and have promoted the shift towards a climate change–centred narrative. As a result of the UN Millennium Summit in 2000, member states adopted the Millennium Development Goals (MDGs), eight broad goals meant to be achieved by 2015. A component of Goal 7, 'Ensure Environmental Sustainability', included improving water sanitation and access to safe drinking water, which meant that governments were responsible for providing this to their citizens (United Nations, 2015). Although India met the MDG water target, sanitation, sustainability and availability of water resources remained problematic areas for the country (United Nations India, 2015). In India's pursuit of achieving MDG targets, more attention was paid towards the river's conservation and cleanliness. In 2008, the National Plan on Climate Change was announced, a major government initiative that included a Ganges conservation project (United Nations India, 2015). It also introduced the NWM, which tackled water security issues with respect to climate change (Ministry of Water Resources, 2008). After the deadline passed in 2015, the Sustainable Development Goals (SDGs) replaced the MDGs. These much broader and more extensive goals set targets to be achieved by 2030. Although one of the targets within Goal 6, 'Water and Sanitation', specifically states to 'implement integrated water resources management at all levels, including through transboundary cooperation as appropriate', six of the total eight targets relate in some form to protection of water resources – be it through sanitation or pollution reduction (United Nations, 2015). This demonstrates that water resource management is generally not a top priority for the SDGs; water security as per the SDGs is dominated by the climate change narrative. Considering that the SDGs are meant to provide direction for sustainable development, it is expected that participating nations follow through on the mandates it outlines. However, only one out of eight targets mentions water management and transboundary cooperation. While each target is intended to be prioritized and fulfilled in concert with all others, it is plausible to imagine a scenario where the development of better water governance is neglected in favour of the seven other targets. Also taking place in 2015, the twenty-first annual UNFCCC Conference of the Parties (COP21) helped shape the current global climate security discourse around INDCs (see Introduction to this volume). Much like past agreements, the COP21 commitments frame water-related issues through a particular vernacular of sovereign states, which may not have a positive effect on transboundary relationships. It is axiomatic that climate change must be recognized for its impact on water accessibility and quality, particularly in states that are known for high carbon outputs and water pollution. However, the use of climate change to frame issues must also be acknowledged for its ability to bury poor management practices under the guise of carbon emissions. In terms of current climate agreements and development plans, the same issue is presenting itself. Scarcity, extreme weather events, and weather variations dominate the language, yet nothing is said about how existing agreements, high modernism, or India's hydro-hegemony may be the underlying factors exacerbating or creating water problems for adjacent states (Warner *et al.*, 2017). As such, global pressures around climate change in multinational forums are propagating the context that will certainly exacerbate negative local-level side effects and which may ultimately lead to boomerang effects within and among basin states. In our view, these climate change scarcity-driven narratives reinforce dominant perspectives that prevent the GBM basin from forging more progressive water policies and practices.

In terms of India's Intended Nationally Determined Contribution (INDC) specifically, there are a number of elements that shed light on how climate change is conceptualized in terms of a problem and potential solutions. From a state security standpoint, the INDC is harkening back to the days of old, where climate change was seen as a challenge to state security in economic and resource terms, not as a problem unto itself. In this sense, the INDC captures this attitude, stating:

> The desire to improve one's lot has been the primary driving force behind human progress. While a few fortunate fellow beings have moved far ahead in this journey of progress, there are many in the world who have been left behind. Nations that are now striving to fulfill this 'right to grow' of their teeming millions cannot be made to feel guilty of their development agenda as they attempt to fulfill this legitimate aspiration. Just because economic development of many countries in the past has come at the cost of environment, it should not be presumed that a reconciliation of the two is not possible.... India, even though not a part of the problem, has been an active and constructive participant in the search for solutions. Even now, when the per capita emissions of many developed countries vary between 7 to 15 metric tonnes, the per capita emissions in India were only about 1.56 metric tonnes in 2010. This is because Indians believe in nature friendly lifestyle and practices rather than its exploitation.
>
> (MOEF Government of India, 2015: 1–2)

By using this language, it is clear that India is resistant to exogenous national pressures, as they could challenge their ability to compete economically on the world stage. This fosters an approach to water management that prioritizes national-scale water issues pertaining to quality and quantity of resource, not poor management. To this point, the NWM focuses exclusively on the issue of climate change and its effect on water availability and supply, waste, or infrastructure (Ministry of Water Affairs, 2008). Meanwhile, in-depth discussion of management issues is absent and left at a mere mention. The NWM emphasizes improved efficiency of water use by at least 20 per cent as its primary goal, which suggests that government priorities lie in tackling water scarcity, but scarcity that has emerged as a result of climate change. While the language used in these documents and India's participation with climate negotiations appear to be a divergence from high-modernism, it is discernible that there is a hesitation to look to current agreements, farming practices, or the approximately 5,000 dams as the main culprits for water problems. In terms of how this affects framing of the Barrage, the NWM notes that the 1996 Ganges Water Treaty did not take into account climate change and is therefore virtually obsolete. 'We need to understand that climate change was not reasonably foreseen at that time …' it says, 'the operation of this general clause may have to consider those circumstances' (Ministry of Water Affairs, 2008: II,53).

Consistent with the country's domestic policy papers around climate change, Bangladesh's INDC looks at how the phenomenon has affected the hydrology of the region and how to mitigate and adapt to climate variations going forward. The document goes so far as to say '… extreme temperatures, erratic rainfall, floods, drought, tropical cyclones, rising sea levels, tidal surges, salinity intrusion and ocean acidification … are gradually offsetting the remarkable socio-economic development gained over the past 30 years, as well as jeopardising future economic growth' (MOEF GPRB, 2015). Further, when Bangladesh outlines its climate mitigation and adaptation priorities, it relegates policy and nationwide capacity building to the last priority. Together these suggest that the current administration does not consider its bilateral agreements with India as the preeminent factor contributing their aforementioned problems. Or, if they do, they may consider them too politicized or securitized to become the object of analysis in any discussions regarding water security.

Lessons learned/moving forward

Despite its political sensitivity, the experience of the Farakka Barrage and climate change illuminates the importance of political economy and framing of water issues. For stakeholders involved with the Farakka Barrage or bilateral water-sharing agreements, it is critical to understand the power relationships of these nations. As a BRICS nation, India has been an active participant in global initiatives around water and the environment, but the pursuit of modernization is still at the forefront of the country's development agenda. The misalignment between India and Bangladesh's perceptions of state, environmental and human security has entrenched the stalemate between these two countries. While both have given increased attention to climate change to improve resiliency, it could be increasing vulnerability by maintaining the same paradigm.

With these problems in mind, the agreements reached at COP21 should be evaluated for potential to mitigate these problems in an inclusive and equitable way. In terms of potential opportunities, India has acknowledged the merits of

Integrated Water Resource Management (IWRM) through a watershed focus; however, there is no tangible plan for how this will come to fruition. The NWM mentions potential avenues through the South Asian Association for Regional Cooperation (SAARC) and/or The Joint River Commission, but this would involve the relinquishment of power over water resources. However, given India's history as the regional hydro-hegemon, it is not clear whether or not the state considers this a viable solution to its own problems or if this is a way of saving face in an international forum. If India is truly concerned about climate change and its effect on water resource, it would integrate more water sharing into its economic development plan as we head towards 2020. While this would improve water resource–management performance in relation to anthropogenic and climate change–based stressors, it would require a rebalancing of national goals. Since the momentum of history is strong, the prudent way forward might involve harvesting low-hanging fruit through a continued focus on improving existing bilateral agreements (Mirumachi, 2015).

In an effort to understand regional intentions moving forward, it is useful to examine the rationale behind ongoing projects, such as Bangladesh's attempts to initiate a bilateral agreement with India to build a new barrage in Bangladesh called the Ganga Barrage. Bangladesh's rationale behind this barrage is that it would eliminate the debilitating salinity issues being inflicted on the southwest region; provide water security throughout the dry season via reservoir storage; and, as a selling point to India, potentially hold enough water that the surplus can be directed down the Hooghly to aid in silt removal at Kolkata Port. Given the analysis above, we are driven to question such persistent piecemeal approaches to water-related challenges. In line with the ongoing problematic management of the Ganges, Bangladesh seems to be pushing a techno-economic (short term) solution that leaves fragmented management practices and high-modern 'man over nature' approaches in place. One doubts, therefore, that the recycling of the barrage idea as both a mitigation and adaptation strategy for water security is anything other than old wine in a new bottle.

In any event, the Ganga Barrage has been met with reservation and push-back from India over concerns of flooding and other potential negative impacts – a scenario where India is protecting itself from damage originating from a project to benefit Bangladesh. This stands in stark contrast to both the Farakka Barrage and the NRLP, which does alter Bangladesh's water availability while solely benefiting India. This juxtaposition in awareness of the negative impacts of poor water management highlights the continued hegemonic dominance India exerts on Bangladesh as the upper riparian state, even as the global climate around water management pushes for transboundary unity and partnerships within water basins.

Considering the desire of both countries to continue modernizing, the nations are balancing their commitments towards global climate initiatives, like the SDGs and COP21, with national pragmatism. The dominant paradigm around climate change adaptation and/or mitigation has attempted to push development in a more progressive direction, but it instead appears to be empowering the region, specifically India, to maintain water management practices rooted in the high-modern era. India's perception of the success of the Farakka Barrage has created fertile ground for a boomerang effect deriving from state-centric perspectives reliant on the construction of large water infrastructure – such as the Ganges Barrage or the controversial Sardar Sarovar project (Mollinga, 2016) – along with a project such as the NWLP. All of these projects attest to the contradictions found within the basin's current problematic ideologies for water management, which desire to support SDG-styled sustainable development but fall short of understanding what that fully entails.

As India continues to modernize and mould itself in our changing global climate, it is important that decision-makers there improve development strategies as well as the principles guiding government policy. Analyzing the Farakka Barrage highlights how India has been an active participant in global initiatives around water and the environment, but has not actually updated core practices or ideologies since the high-modern era. It is undeniable that for India's progression as a nation they need to see past the excuses afforded to them by modern climate change discourse and recognize that their actions, beyond carbon emissions, greatly impact water availability.

References

Allen, T. (2003, April). IWRM/IWRAM: a new sanctioned discourse. *SOAS Water Issues Study Group, Occasional Paper No. 50.*, 1–27.

Amarasinghe, U. (2009, August). Strategic Analysis of India's River Linking Project. *CGIAR Challenge Program on Water and Food Project Report, Project No. 48.*, 1–76.

Bandyopadhyay, J., and Pervenn, S. (2003, June). The interlinking of Indian rivers: Some questions on the scientific, economic and environmental dimensions of the proposal. *SOAS Water Issues Study Group, Occasional Paper No. 60*, 1–34.

Conaway, C. (2015, 23 September). The Ganges River is dying under the weight of modern India. *Newsweek.* Accessed 5 April 2016. www.newsweek.com/2015/10/02/ganges-river-dying-under-weight-modern-india-375347.html.

Das, P., and Tamminga, K. R. (2012). The Ganges and the GAP: An assessment of efforts to clean a sacred river. *Sustainability*, 4, 1647–1668.

Earle, A., Cascao, A. E., Hansson, S., Jägerskog A., Swain, A. and Öjendal, J. (2015). *Transboundary Water Management and the Climate Change Debate.* London: Routledge.

Gain, A. K., and Giupponi, C. (2014). Impact of the Farakka dam on thresholds of the hydrologic flow regime in the lower Ganges river basin (Bangladesh). *Water*, 6(8), 2501–2518.

Hossain, M. M., Zaman, A. M., and Ludwig, F. (2015). Climate change impact on the discharge of Ganges-Brahmaputra-Meghna (GBM) River Basin and Bangladesh. Paper presented at the International Conference on Climate Change in relation to Water and Environment (I3CWE-2015), Gazipur, Bangladesh.

Hossen, M. A. (2015). The Ganges Basin management and community empowerment. *Bandung Journal of the Global South*, 2(1).

IWMI. (n.d.). Strategic analysis of India's national river-linking project. International Water Management Institute. Accessed 10 April 2016. http://nrlp.iwmi.org/main/maps.asp#

Kawser, M., and Samad, A. (2016). Political history of Farakka Barrage and its effects on environment in Bangladesh. *Bandung: Journal of the Global South*, 2(1), 1–14.

Kumar-Rao, A. (2015, June 2). Photo essay: The nowhere people. *The Third Pole*. Accessed 9 April 2016. www.thethirdpole.net/2015/06/02/photo-essay-the-nowhere-people/

Meakin, S. (1992, November). The Rio Earth Summit: Summary of the United Nations conference on Environment and Development (BP-317E). Retrieved 5 April 2016, from http://publications.gc.ca/Collection-R/LoPBdP/BP/bp317-e.htm#B. Countries(txt)

Ministry of Environment and Forests (2009). Bangladesh climate strategy and action plan. Working paper. Dhaka, Bangladesh: Government of Bangladesh.

Ministry of Environment and Forests (MOEF) Government of India. (2015). India's Intended Nationally Determined Contribution: Working towards climate justice. Retrieved 1 April 2016, from www4.unfccc.int/submissions/INDC/Published Documents/India/1/INDIA INDC TO UNFCCC.pdf

Ministry of Environment and Forests (MOEF), Government of the People's Republic of Bangladesh (2012). Rio + 20: National Report on Sustainable Development. Retrieved 15 May 2018, from https://policy.asiapacificenergy.org/sites/default/files/Rio%2B20_Bangladesh_reduced.pdf

Ministry of Environment and Forests (MOEF), Government of the People's Republic of Bangladesh (GFRB) (2015). Intended Nationally Determined Contributions (INDC). Retrieved 1 April 2016, from www4.unfccc.int/submissions/INDC/PublishedDocuments/Bangladesh/1/INDC_2015_of_Bangladesh.pdf

Ministry of Water Affairs. Government of India (2008). National Water Mission under National Action Plan on Climate Change. Comprehensive Mission Document Volume II. New Delhi: Government of India.

Mirumachi, N. (2015). *Transboundary Water Politics in the Developing World*. London: Routledge.

Mollinga, P. (2016, 25 February). Downstream of the Dam: Farmers, pipelines and capitalist development in the Sardar Sarovar project. Brunei Gallery, SOAS Inaugural Lecture Series, (London, UK).

Moudud, H. J. (1998). Ganges water dispute: A case study of people's participation and public awareness. *Coastal Area Resource Development and Management Association*. Retrieved 1 April 2016.

Rahaman, M. M. (2006). The Ganges water conflict. *International Water Law Project*. www.internationalwaterlaw.org/bibliography/articles/general/Rahaman-Ganges-Asteriskos.pdf

Salman, M. A., and Uprety, K. (2002). Conflict and cooperation on South Asia's international rivers: A legal perspective. *World Bank Publications*.

Swain, A. (1996). Displacing the conflict: environmental destruction in Bangladesh and ethnic conflict in India. *Journal of Peace Research*, 33(2), 189–204.

Thakur, P., Laha, C., and Aggarwal, S. P. (2012). River bank erosion hazard study of river Ganga, upstream of Farakka barrage using remote sensing and GIS. *Natural Hazards*, 61(3), 967–987.

Thomas, B. K., Jamawl, P., Lele, S., and Srinivasan, V. (2014). Thinking about urban resilience: The case of water scarcity and wastewater reuse in Bengaluru. *Urban Resilience: Proceedings of the Colloquium* (pp. 34–37). Bengaluru, India: Environmental Governance Group.

United Nations. (23 May 1997). Earth Summit: UN Conference on Environment and Development. Retrieved April 1, 2016, from www.un.org/geninfo/bp/enviro.html

United Nations. (6 July 2015). The Millennium Development Goals Report 2015. Accessed 12 April 2016. www.undp.org/content/dam/undp/library/MDG/english/ UNDP_MDG_Report_2015.pdf

United Nations India (2015). India and the MDGs. February 2015. Accessed 12 April 2016. http://in.one.un.org/img/uploads/India_and_the_MDGs.pdf

Warner, J., Mirumachi, N., Farnum, R. L., Grandi, M., Menga, F., and Zeitoun, M. (2017). Transboundary 'hydro-hegemony: 10 years later. *WIREs Water*, 4(6), e1242.

Wolf, A. T., and Newton, J. T. (2007). Case study of transboundary dispute resolution: The Ganges River controversy. Oregon State University: Transboundary Freshwater Dispute Database. www.transboundarywaters.orst.edu/research/case_studies/Ganges_ New. htm

Zeitoun, M., Lankford, B., Krueger, T., Forsyth, T., Carter, R., Hoekstra, A. Y., Taylor, R., Varis, O., Cleaver, O., Boelens, R., Swatuk, L. A. Tickner, D.,. Scott, C. A., Mirumachi, N., and Matthews, N. (2016). Reductionist and integrative research approaches to complex water security policy challenges. *Global Environmental Change*, 39, 143–154.

5 Contested development
The Belo Monte Dam, Brazil

Jason Durst, Liam Neumann, Anna Smith,
Luis Paulo Batista da Silva and Larry Swatuk

Introduction

Brazil is the world's fifth largest country, with a population of 200 million, and has the seventh largest economy (BMI, 2016). The country is also forecast to see high rates of future growth across the socioeconomic spectrum (BMI, 2016). For instance, energy demand, a proxy for broader socioeconomic development, is projected to double in a decade. Meeting future energy demands, therefore, will be central to the development challenges faced by Brazil over the coming years. According to the International Energy Agency (IEA) (2016a), the Brazilian government has correspondingly pledged to double energy production by 2024. Hydropower production will remain the dominant form of energy production, comprising over 70 per cent of Brazil's total energy production, meaning that installed capacity will have to increase by a considerable degree. The largest hydroelectric facility currently under construction is the Belo Monte Dam on the Xingu River, a major tributary of the Amazon River. When completed, Belo Monte will be the third largest hydroelectric dam in the world, providing enough energy to power 20 million Brazilian homes (see Figure 5.1).

This chapter uses the Belo Monte Dam as a case study to examine the challenges for development that can arise when policies are made without sufficient regard for localized contextual factors in the development process. Furthermore, utilizing the conceptual framework of the boomerang effect, the discussion to follow will critically examine the narratives surrounding Belo Monte via three distinct sections. The first examines Belo Monte within the context of Brazilian energy policy, and as a representation of hydroelectricity more generally within Brazil. The second, will provide a detailed stakeholder analysis of Belo Monte. Stakeholders are divided into those who are projected to benefit from Belo Monte, those occupying a middle ground, and those who are projected to be negatively impacted by the dam. The final section examines how the narrative surrounding Belo Monte has changed over time, from the project's original inception in 1975 until the present day. In conclusion, it is determined that although projects such as Belo Monte could have significant benefits when considered in isolation, it is essential to take a holistic perspective when determining the true costs and benefits of such development projects.

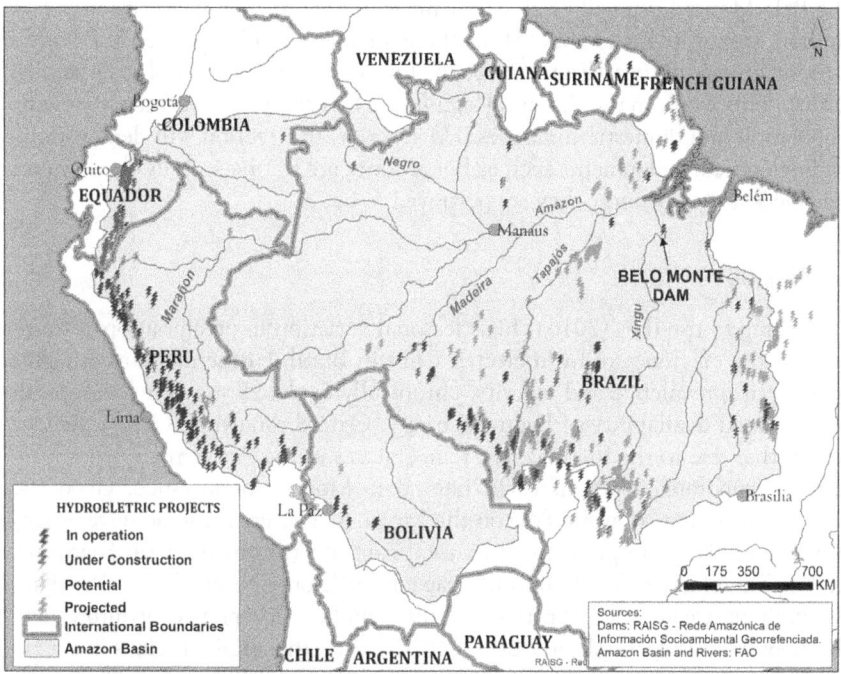

Figure 5.1 Map of Belo Monte Dam in relation to other dams in South America.

Belo Monte and hydroelectricity in Brazil

By 2024 Brazil will see a population increase of approximately 20 million, and according to Business Monitor International (BMI) (2016) they will be twice as rich as they are today. This will result in a significant increase in total energy demand. Moreira *et al.* (2015) and Bratman (2014) project that Brazilian domestic energy consumption will grow by around 3–5 per cent annually up to about 2025, outpacing both population and economic growth. Meeting future energy demand is therefore a central focus of the Brazilian government. In recent COP21 talks in Paris, Brazil committed to a 37 per cent reduction in emissions from 2005 levels by 2025 (British Broadcasting Corporation, 2010. Hydroelectricity has therefore been portrayed by the government as a low-cost source of reliable clean energy. A Swedish Agency for Growth Policy Analysis Report (2013) shows that future government targets maintain the current dominance of hydroelectric generation while aiming to phase out use of biomass and coal. According to BMI (2016), the hydropower sector will see the biggest expansion in terms of new installed capacity over the coming decade. According to the proponents of the Belo Monte Dam, hydroelectricity is an affordable and reliable source of energy that will allow Brazil to meet its future energy demands in an environmentally sustainable manner (Bratman, 2014; Petras,

2013). For instance, through the use of the natural elevation of the Xingu River the Belo Monte Dam will be able to achieve a theoretical annual capacity of 11 million megawatt hours (MWh), or the equivalent of 12.5 per cent of Brazil's current energy output (Bratman, 2016; International Rivers, 2016; Fearnside, 2006). This would naturally bring significant benefits to the country's growing population and domestic industries. However, as this section will demonstrate, hydroelectric developments such as Belo Monte are not necessarily the low cost reliable sources of energy they are portrayed to be.

Not reliable

According to the IEA (2016a) high seasonal variation in precipitation patterns reduces the efficiency of hydroelectric dams in Brazil. Furthermore, as droughts increase in prevalence and severity, chronically depleted reservoir levels result in decreased availability of hydroelectricity. Oxford Analytica (2014) demonstrates that due to the high average temperatures in the Amazonian region, reservoirs evaporate at an increased rate. Belo Monte, for instance, given the seasonal variations in water flow on the Xingu, will be unable to achieve energy production above 30 per cent of its total capacity (Bratman, 2014; Fearnside, 2006). Therefore, Belo Monte's full capacity will only be viable if other dams are built upstream – its construction thus implicit acceptance of future dam development projects (Fearnside, 2016. Indeed, the original plans for Belo Monte comprised five additional up-stream dams with significantly higher environmental and sociocultural impacts (International Rivers 2016; Jaichand and Sampaio, 2013). Moreira *et al.* (2015) illustrates the relative and absolute unreliability of hydroelectricity in Brazil: relative as compared to nuclear, biomass and fossil fuel based electricity generation; absolute in that hydropower's reliability, measured on an index of 0 to 1, with 1 being perfect reliability, is slightly more than 0.2.[1] Water shortages in 2001, for instance, resulted in a crisis for the electricity sector. With depleting reservoirs, the major municipalities of São Paulo, Rio de Janeiro, and Minas Gerais – together responsible for half of Brazilian GDP – were forced to introduce a rationing plan for electricity (Meisen, 2010). To accommodate these challenges Brazilian governments have traditionally depended upon the use of natural gas.

Not cheap

Hydroelectricity has historically dominated electricity supply in Brazil. However, remaining hydroelectric potential is mainly located far from the centres of population and industry (Bratman, 2014). Brazil's remaining hydroelectric potential is located both far from the existing transmission grid and from major centres of population. The IEA (2016b) and World Bank (2016) also state that due to Brazil's hydropower generating facilities (in the central and northwest) being located far from the main demand centres (along the coasts and in the south east), there exists high transmission and distribution losses. Belo Monte, for example, is located over 2,000 km from the major population centres of Rio and Sao Paulo where a third of the electricity generated at the dam will be used. Such distances, according to the IEA (2006), are a primary contributor to Brazil's high rates of losses during distribution, which at roughly 17 per cent of total generation is one of the highest in the world.[2] Therefore, despite the generally low costs associated with hydroelectric power generation, prices for final consumers in Brazil can exceed the costs of generation by as much as 40 per cent (IEA, 2006). Expanding hydroelectricity further into the remote northern regions will therefore only contribute to further losses and inefficiencies within the energy system.

Hydroelectric developments have also been shown to have the potential for significant environmental impacts, particularly when occurring in ecologically and socioculturally sensitive areas. The Brazilian Amazon, for instance, has one of the highest concentrations of biodiversity on earth, and is home to a third of the world's rainforests (Bratman, 2015). Research conducted by Tundisi (2014) and Moreira *et al.* (2015) shows that hydroelectricity is second only to biomass in terms of direct impact on the environment, primarily through direct changes to the landscape. Whereas supporters of the Belo Monte project argue that its relatively small reservoir size limits its impact on the environment, its critics argue that the natural flow of the Xingu will be permanently altered.[3] According to Benchimol (2015) up to 80 per cent of the downstream Xingu will be diverted for energy production. This reduction of the downstream hydraulic cycle could result in substantial losses of aquatic and terrestrial fauna, significantly increasing the possibility of species extinction (Benchimol, 2015; Jaichand and Sampaio, 2013). Furthermore, according to Diamond (Diamon and Poirier, 2010) and Fearnside (2006) the Xingu River has an extraordinary diversity of indigenous cultures, and the areas affected by the Belo Monte Dam are currently home to 37 distinct ethnicities. Bratman (2015) claims that a minimum of 19,242 people, most of whom are indigenous, will be displaced as a result of Belo Monte. This number, however, will likely be much higher as over 100,000 people live within the region directly impacted by the dam (Zortéa *et al.*, 2015; Bratman, 2014). As such, the Belo Monte project has been the subject of numerous protests and public lawsuits. Indeed, the Belo Monte Dam demonstrates the challenges faced by Brazilian hydroelectricity in general. As the next section will demonstrate, the impacts of Belo Monte can be disaggregated into stakeholder groups of those who will benefit, those who will not, and those in the middle.

Stakeholder analysis

This section provides a detailed stakeholder analysis of Belo Monte. Stakeholders are divided into groups representing actors benefiting from Belo Monte, those occupying a middle ground, and actors projected to be negatively impacted by the development of the Belo Monte Dam (see Table 5.1 below).

Table 5.1 Stakeholders of the Xingu River Basin in relation to the Belo Monte project

Actors	Influence	Interest	Position in 2000	Position in 2016
Central Government	Primary decision-making power	Economic growth; employment creation; access to electricity; social responsibility; bribery and incentives; political support	For the Dam	Delayed the Dam
Brazilian National Development Bank (BNDES)	Secondary decision-making power	Economic growth; access to markets	For the Dam	No change (n/c)
Private Investors	Secondary decision-making power	Economic growth; access to markets	For the Dam	n/c
Electrobras (state-owned company)	Primary decision-making power	Economic growth; access to markets; employment creation	For the Dam	n/c
Ministry of Development, Industry and Foreign Trade	Secondary decision-making power	Economic growth; access to markets; employment creation; maintain budgetary allocations; service delivery	For the Dam	n/c
Norsk Hydro (mining company)	Secondary decision-making power	Access to energy allows for economic growth	For the Dam	n/c
São Paulo and Rio De Janeiro	Secondary decision-making power	Access to energy; Social rights; Environment protection	For the Dam	Against the Dam
Ministry of Environment (IBAMA)	Primary decision making power	Environmental sustainability; employment creation; sustainable water supply; quality control; contribute to national development; social welfare	Debated	Delayed the Dam
Researcher (i.e. Universidade Federal do Para)	Knowledge-based power	Funding, credibility, environmental and social protection	Debated	n/c
Indigenous people	Social movement-based power	Social welfare; human rights; environmental health and sustainability	Against the Dam	n/c
Local communities and small cities effected by the dam	Social movement-based power	Social welfare; human rights; environmental health and sustainability; access to energy	Against the Dam	n/c
NGOs	Secondary decision-making power	Human rights and dignity; environmental health and sustainability	Against the Dam	n/c

Those who will benefit

The stakeholders who are providing the financing for Belo Monte are projected to be the ultimate beneficiaries of the project. Financing for Belo Monte is provided by the Brazilian Development Bank (BNDES), which accounts for 80 per cent of total investment, with a consortium of private investors (including construction and energy companies) comprising the remaining investment portfolio (Bratman, 2014). However, the BNDES has also received a 500 million loan for projects supporting mitigation and energy efficiency for Brazil from the European Investment Bank, partially making this an international matter (Steffek and Romeiro 2016). The BNDES's role as a principal financer of Belo Monte increases the Bank's responsibility for the project's environmental and social effects, thus elaborate social responsibility initiatives such as the Amazon Fund, the Climate Fund Program as well as several other environmental and social funds have been developed (BNDES, 2016), presumably attempting to reduce public pressure. Norte Energia, a consortium made up of state and private companies, is considered owner of the Belo Monte Dam (Norte Energia, 2016). The government has a majority ownership of the dam at approximately 49 per cent through Electrobras, a state-owned company, as well as their two subsidiary companies Electronorte and CHESF (Guatam *et al.*, 2014). The other 51 per cent of owners includes a few investment funds and multiple construction companies (Bratman, 2015). The cost of the dam is an estimated $30 billion including transmission lines, yet it is predicted generation of the dam will exceed these costs (Diamond *et al.*, 2010).

Another major benefactor of Belo Monte will be the mining and extractive industries of Brazil. According to Westholm *et al.* (2011) mines require a high amount of energy to operate, and are furthermore generally located in remote locations. Brazil currently allocates large portions of the energy grid to subsidized aluminium smelting for export, and according to Bratman (2014) these industries will likely be the greatest benefactors of increased hydro capacity. The government of Brazil will also benefit from Belo Monte. The Ministry of Development, Industry and Foreign Trade, for instance, receives levies from industrial activities and therefore will indirectly receive benefits from the dam's completion. Government officials in Brazil are also able to personally benefit from such projects. Recent corruption scandals at state-owned Petrobras have shown the endemic nature of corruption in Brazil. For instance, Camargo Corrêa, a large scale construction company that holds a 16 per cent share of the multibillion dollar Belo Monte construction cost, paid the Partido dos Trabalhadores and Partido do Movimento Democrático Brasileiro parties over $100 million in bribes (Bratman, 2016). Furthermore, according to Bratman (2015), another half dozen construction companies are under investigation for corruption under similar circumstances.

Caught in the middle

The Ministry of Environment is stuck between two demands. On the one hand, indigenous groups, NGOs and civil society come to the Brazilian Institute of Environment and Renewable Natural Resources (IBAMA) claiming that they must protect the environment. On the other side are the ministries of the government and the central government itself pressuring IBAMA to provide approval for the project. Indeed, IBAMA gave into government pressure by providing Electrobras with provisional environmental licenses to develop the dam in 2000 and 2010 (Zanotti, 2015). However, in doing so they have lost the legitimacy and respect of the people. Due to the controversy, two directors and a president have quit over the process (Baptista and Thorkildsen, 2011). Researchers involved in the dam are Universidade Federal do Para (UFP), Nucleus for High-Level Studies of Amazonia (NAEA), the Emílio Goeldi Museum of Para' (MPEG) and the National Institute for Research in the Amazon (INPA). Much of the research completed is questionable due to funding bias. For example, Electronorte funded UFP $1.3 million for the Belo Monte research project, which account for one-fifth of the university's entire budget (Berry and Mollard, 2010. Other approaches also question fraudulent data. During the research of the dam it was claimed that Electronorte's staff manipulated data and attempted to control researches through monitoring and advising researchers in the field (Berry and Mollard, 2010). Electronorte reportedly even asked for the raw data before the studies had been completed (Berry and Mollard, 2010). Researchers on the project are torn between legitimacy and funding. Finally, although initially the people living in São Paulo and Rio de Janeiro considered themselves to be beneficiaries through blackout reductions (Fearnside, 2006), it has become increasingly evident that simply increasing the amount of hydroelectric generation will not solve Brazil's problems, thereby shifting public support towards the local communities who oppose the dam (Oliver Smith, 2014).

Those who will not benefit

This subsection assesses a multitude of stakeholders that have opposed and are negatively affected by the construction and operation of the dam. The main threat of the dam is the reality that people will lose their livelihoods. Flooding accounts for the destruction of homes of 40,000–60,000 indigenous people, land owners in the region and people in the city of Altamira (Randell, 2015). There are also 800 people from the Juruna, Xikrin, Arara, Xipaia, Kuruaya, Kayapó and other indigenous groups that will be displaced from their livelihoods such as fishing and trading along the river. New compensation provided by the government and Electronorte have developed homes for the indigenous. However, negative effects have already sprung from these communities, such as soaring cholesterol rates, minimal sanitation and high electricity bills (Amazon Watch, 2016). Furthermore, the loss of their ancient practices forces alterations to entrenched cultural identities as they must now adopt alternative food production methods. This has also led to a major infestation of mosquitoes in the newly built communities.

Other stakeholders being excluded from the benefits of the dam are close local communities that are energy insecure. These communities have not been given priority in receiving any energy from the dam, or at a discounted rate (Bratman, 2015). Some of these communities are São Antonio and Altimira. Altimira, an estimated population of 100,000, will have about 20,000 displaced due to the dam (Bratman, 2015). Social movements have developed over the years in response to the Belo Monte Dam, such as the Movement of People Affected by Dams (MAB). They are Brazil's second-largest social movement, protecting those being displaced due to industrial public and private energy companies. Other important NGOs involved are the Xingu Alive Forever Movement, and the Movement for the Development of the Trans-Amazon and Xingu (MDTX) (Bratman, 2015). Also included are international organizations such as International Rivers, UN Human Rights Council and the International Labour Organization. They stated that the Brazilian government has violated the ILO convention on the protection and human rights of indigenous and tribal people (Jaichand and Sampaio, 2013). Local and international NGOs are fighting for the people and ecosystems negatively affected by the dam; however, the Brazilian government has largely disregarded these outcries. While recent Brazilian government claims to reconsider its commitment to mega-dams and other large scale projects are viewed by some as indicative of successful activism on the part of those most negatively affected, others are not so charitable in their interpretation (Watts, 2016).

Not only will the dam affect people locally, but will reach globally. The project and process of the dam will emit a high amount of CO_2 emissions as well as contribute to biodiversity loss (Benchimol and Peres, 2015). This pollution affects our planet in terms of climate change and therefore affects us all.

The narratives surrounding Belo Monte

This section will examine how the narratives surrounding Belo Monte have transformed over time as a result of factors at both the local and international level. Belo Monte had been enlaced in a rhetoric of modernization, which shifted into a framing that aligned with an acceptance of responsibility for the natural world, and, finally, contemporary amalgamations of past trends.

Modernization

The first proposal for the Belo Monte Dam occurred under a military dictatorship in 1975. The reservoir as then conceived was considerably larger than what is currently under construction (Bratman, 2014; Castro, 2004). According to Bratman (2014) the narrative in 1975 was framed around economic growth and infrastructure building – with the Amazon as a potential resource to be exploited. Dam construction was thus regarded as intrinsic to the government's vision of modernization and essential for growth (Bratman, 2015). The massive developments that such a narrative inspired were often implemented without any form of public or environmental considerations. Indeed, according to Fearnside (2009) the technocratic elite determining these policies considered them to be essential for the country's progress. However, increased knowledge of the environmental and social consequences of the hydroelectric developments in the Amazon was built up in response to these policies, and according to Bratman (2014) has been a major contributor to the development of civil society within Brazil. Indeed, more than half a dozen studies were conducted on the early Belo Monte proposal to assess the potential social and environmental impacts (Jaichand and Sampaio, 2013). These studies demonstrated that the negative effects on the environment and nearby settlements would be significant. In 1989, indigenous groups and members of international civil society organizations rallied at Altamira to protest against the planned hydroelectric power scheme on the Xingu. According to Baptista and Thorkildsen. (2011), celebrities such as Sting and the founder of the Body Shop, Anita Roddick, participated in an increasingly international lobby against Belo Monte. The Altamira gathering resulted in the World Bank's cancellation of a loan in support of the Belo Monte project and, according to Oliver-Smith (2014), was primarily responsible for the decade-long halting of construction.

Socio-environmental responsibilities

Brazil's regime, which had disregarded the social and environmental impacts of their growth strategy, did not fare well in moving forward with their plans. The process to which they had ascribed disallowed their initiatives being fulfilled and played a role in the establishment of polarized ideologies (Bratman, 2015). On one side resided those ascribing foremost to capitalist and state interest, inclusive of the government; on the other, the anti-dam proponents arguing for environmental and human rights (Bratman, 2016). In response to the barriers faced in constructing the dam, the government has skilfully managed to reinvent Belo Monte, and its own image, in order to jump through the hoops necessary to reach its desired objectives (Bratman, 2015).

The Belo Monte project was halted in 1989 for a number of reasons. For instance, in 1988, a global campaign painting the Brazilian government as the perpetrator in destroying the Amazon due to released deforestation figures was organized, and gained significant international attention (Fernandes, 1992). The increasing pressures facing the government to place greater significance on environmental protectionist measures undoubtedly persuaded the Constituent Assembly to accept the Brazilian Federal Constitution of 1988, which had more environmental safeguards than Brazil had previously seen (Fernandes, 1992). Furthermore, according to Article 231 of the Constitution, indigenous consent is required for any projects with impacts on indigenous lands (Fearnside, 2006).

By removing Belo Monte from the spotlight, the Brazilian government was therefore able to reframe the narrative around engagement with public concerns for the environment and human rights. While environmentalists, social movements and the actions of the local population constrained the government's ability to move forward with its planned series of dams on the Xingu River, Fainguelernt (2016) points out that the collapse of the military-led regime under the 1980s debt crisis, and the emergence of the redemocratization project under a new constitution, was equally important in restraining such profligate energy works.

In the run-up to the 1992 Rio de Janeiro-hosted world summit on environment and development (the so-called 'Earth Summit'), the Brazilian government worked to reconfigure Belo Monte within the framework of climate change and environmental protection. Given global pressure regarding environmental sustainability and internal pressures related to renewed economic growth, the Brazilian government changed Belo Monte's project to make it feasible. In order to rebrand Belo Monte for approval the name was changed from Kararao to the current Belo Monte in 1998 (Carvalho, 2006). Instead of five dams, the government made assurances that only Belo Monte would be carried out within the Xingu basin. Given the technological developments since the 1980s, the flooded area needed to run the turbines decreased from $1,225 \, km^2$ to $516 \, km^2$ (Fearnside, 2006; EPE, 2011). Moreover, according to the former Minister of Environment (MMA), Carlos Minc, the construction of Belo Monte would reduce the participation of thermoelectric plants on Brazilian energy matrix. Therefore, according to Brazilian government, Belo Monte became a tool to fulfil its climate goal to reduce CO_2 emissions until 2020 (Gramacho, 2010).

The current state of affairs

Belo Monte was once again up for official approval in 2005, with much contention in the years to follow. This time, however, strong supporters of the dam would take to the streets – symbolizing the increase in the government's allies for the project (Bratman, 2015). Community protest would prove to be less effective, though similar tactics were used in the organized 2008 meeting to those of 1989 (Bratman, 2014). After 2010, the plans put forward by Electronorte, who had won out the bidding to build Belo Monte, added further alterations to the plans (Crones, 2014). The new plans would bypass the obligation of granted approval for engaging in activities that directly affect indigenous land, as flooding was completely diverted away from these areas (Bratman, 2015; Crones, 2014; Jaicond and Sampio, 2013). This allowed for plans to legally proceed without having to consult the indigenous population who, although not directly affected any longer, would feel the effects of the river's depleted flow (Crones, 2014). Bratman (2015) argues that the Green Economy discourse has become a hegemonic instrument, as it has effectively undermined opposition of the dam, based on environmental detriment claims, and has helped reposition

Belo Monte as environmentally responsible through its clean energy rhetoric. Indeed, the government's recent defence of the dam has been framed in terms of the necessity to meet future increases in electricity demand (Bratman, 2016; EIA, 2015; Luomi, 2014). This framework presents electrical expansion as an unavoidable phenomenon, and, coupled with its green argument, acts to further legitimize Belo Monte's construction (Graeff, 2012). Further to this, the Rouseff administration attempted to frame Belo Monte not only as a socially and environmentally responsible endeavour, but also one that will bring livelihood improvements. Electrobras,[4] the state-owned enterprise (SOE) that controls the majority of Brazil's electricity sector, does not, for instance, deny that Belo Monte will have significant impacts on the natural and human environment (Norte Energia, 2016). Such concerns, however, are typically accounted for in egalitarian calculations such as cost-benefit analysis (CBA), or impact assessments. By taking into account compensation packages to resettled populations such normative calculations are able to justify the location-specific impacts of Belo Monte by claiming to promote the common good. According to Norte Energia (2016) reallocation can be positive if resettlement occurs in locations with improved sanitation, amenities and public facilities (Norte Energia, 2016). However, Norte Energia is neglecting the impacts that altering a population's homeland can have upon the sociocultural fabric of their society. Indeed, according to Watts (2016) and Grove (2014) recent court rulings in Brazil have found that indigenous communities are systemically undercompensated for the impacts to their communities from developments such as Belo Monte. This is a cyclical dynamic of the government – promoting large hydroelectric developments, succumbing to public pressures against them, only to reframe hydroelectricity as an essential tool for development.

In 2010, in order to have its installation license issued and start to build the dam, NE committed to undertake more than 40 mitigation and compensation actions regarding affected municipalities. The cost of these actions were foreseen as R\$3.2 billion (around US\$2 billion). Accomplishing those actions were mandatory to have Belo Monte's operation allowed. Water provision and sanitation, urban and rural resettlements, education and health infrastructure, public security and ecological compensation were the main themes addressed by public and private studies and agreed by NE through legally binding contracts. A special set of actions was planned to mitigate the impacts of Belo Monte on indigenous people living nearby the dam. Actions on indigenous lands were planned to restructure their livelihoods in the long term, given the new environmental conditions; however, NE has been acting in a patronaging manner, supplying boats, fuel and a two-year monthly allowance to each local tribe (ISA, 2015).

Even though the committed actions were not realized, the Brazilian Institute for the Environment and Renewable Natural Resources (IBAMA) issued Belo Monte's operation license in November of 2015. Since the beginning of the dam's installation, criminality rates have increased in all municipalities affected by the dam, such as murders, rapes and traffic fatalities (Reis and Souza, 2016).

Most of the evacuated urban and rural households have not been properly resettled, either due to insufficient compensation allowances or poor conditions of dwelling units (ISA, 2015). Hospitals and schools are overcrowded thereby increasing time for services in hospitals and decreasing quality in public education. Hence, civil society has been demanding the fulfilment of NE's committed actions and public prosecutors are issuing lawsuits and fines in order to guarantee the mitigation of Belo Monte's impact.

During April of 2017 a Special Court in Brasília, Brazil, ruled a decision ordering the stopping of the operation of six turbines currently working at Belo Monte hydropower dam (Pontes, 2017). Further, in September of the same year, the same court issued another order to stop the works carried out to finish the total of 18 turbines predicted at Belo Monte's project (Harari, 2017). These two judicial decisions are the outcome of several law suits filed by public prosecutors, environmentalists and associations opposed to the plant. The first decision was filed regarding non-fulfilment to undertake sanitation works by NE. The second decision was upheld in order to punish NE for the poor household conditions given to 3,000 impacted families resettled after the reservoir flooding. Despite these decisions, Belo Monte is still operating and the works are still being executed, due to court proceedings and the slowness of judicial enforcement (ISA, 2015; Harari, 2017).

Belo Monte is already a reality in the Brazilian Amazon region. Several cities are affected, as are indigenous tribes and very complex ecosystems. The mitigation and compensation actions were not properly executed, thus Belo Monte's hydropower project cannot be currently assessed as a better practice project. Actually, not even the goal to supply Brazil with green energy in a reliable manner is accomplished. Hence, in relation to the boomerang effect, Belo Monte raises important questions regarding both local-level side effects and feedbacks to the state.

Conclusion

As this chapter has demonstrated it is important to take a holistic perspective when determining the costs and benefits of development projects. Hydroelectric projects can indeed provide important benefits as a source of energy. However, as we have shown, it is important to consider the localized impacts of such projects, as well as the narrative being used to support them. From the stakeholder analysis presented above it can be seen that determining who will benefit from, or be negatively impacted by, a particular project is a complicated process. It is essential, therefore, that potential negative local-level side effects, as well as state-level boomerang effects, be considered when determining policies for development.

Notes

1 A comparison of the reliability of energy sources in the specific context of Brazilian electricity generation reveals the following: biomass 0.6; coal 0.65; wind <0.1; natural gas 0.7; oil 0.5; nuclear 0.8; and hydro 0.21 (Moreira *et al.*, 2015). According to Moreira *et al.* (2015), energy generation reliability is determined through a combination of two factors: availability and capacity. Availability is defined as the amount of time that each resource is able to provide energy on an annual basis. Capacity is the average amount of electricity generated over a year. To generate fully reliable power, or a score of 1, a power plant would need to be continuously available while generating at full capacity.
2 The average distribution losses for the Organisation for Economic Co-operation and Development (OECD), a club of mostly rich nations (but also countries such as Mexico) is 7 per cent. Canada, for instance, has had losses above 10 per cent only once since the 1970s, at 11.4 per cent in 2009 (World Bank, 2016; IEA, 2006).
3 Belo Monte is being built at a portion of the Xingu River with an 80m drop in elevation (high by Amazonian standards). According to Fearnside (2006), utilizing the natural flow of the Xingu River provides the developers of Belo Monte with the ability to claim a high installed capacity without the need for a large reservoir. A relatively small 500 km² area of land will be flooded to create the Belo Monte reservoir.
4 Electrobras owns a majority stake in Norte Energia, which in turn owns a controlling stake in the Belo Monte Dam (Norte Energia, 2016).

References

Amazon Watch. (2016). The high price of Belo Monte and modernization in the life of the Muratu indigenous community. Available at: http://amazonwatch.org/news/2016/0209-the-high-price-of-belo-monte-andmodernization-in-the-life-of-the-muratu-indigenous-community. Accessed 1 April 2018.

Baptista, F. M., and Thorkildsen, K. (2011). The Belo Monte Dam: a camel in the tent?. *Norwegian Latin America Research Network*. Available at: http://search.proquest.com.proxy.lib.uwaterloo.ca/docview/1661346420?accountid=14906. Accessed 1 April 2018.

Benchimol, M., and Peres, C. (2015). Widespread forest vertebrate extinctions induced by a mega hydroelectric dam in lowland Amazonia. *PLoS ONE*, 10(7), 1–15.

Berry, K. A., and Mollard, E., eds (2010). *Social Participation in Water Governance and Management: Critical and Global Perspectives*. London: Earthscan.

Bratman, E. (2014). Contradictions of green development: Human rights and environmental norms in light of Belo Monte Dam activism. *Journal of Latin American Studies*, 46(21), 261–289.

Bratman, E. (2015). Passive revolution in the green economy: Activism and the Belo Monte Dam. *International Environmental Agreements: Politics, Law and Economics*, 15(61), 1–17.

Bratman, E. (2016). Contradictions of sustainable development: Brazil's New Left and Amazonian environmental governance. Paper presented at the annual meeting of the International Studies Association (March).

Brazilian Development Bank (BNDES) (2016). *Institutional Social and Environmental Responsibility in Brazil*. Available at: www.bndes.gov.br/SiteBNDES/bndes/bndes_en/Institucional/Social_and_Environmental_Responsibility/. Accessed 1 April 2018.

British Broadcasting Corporation (BBC) (2010). Brazil Awards Rights to Develop Belo Monte Dam. *The British Broadcasting Corporation*. Available at: http://news.bbc.co.uk/2/hi/8633786.stm. Accessed 1 April 2018.

Business Monitor International (BMI) (2016). *Brazil power report – Q1 2016*. London: Business Monitor International.

Carvalho, G. O. (2006). Environmental resistance and the politics of energy development in the Brazilian Amazon. *The Journal of Environment and Development*, 15(3), 245–268.

Castro, E. M. (2004). Water without dams: Women organizing in the Amazon region. *Women and Environments International*, 64/65, 9–11.

Crones, R. (2014). Hydroelectric Amazonia: Cultures caught in the crucible of order of progress. *Tipiti: Journal of the Society for the Anthropology of Lowland South America*, 12(2), 105–112.

Diamond, S., and Poirier, C. (2010). Brazil's native peoples and the Belo Monte Dam: A case study. *NACLA Report on the Americas*, 43(5), 25–70.

Energy Information Administration of the United States (EIA) (2016). Brazil: An overview. The Government of the United States. Available at: www.eia.gov/beta/international/analysis.cfm?iso=BRA Accessed 1 April 2018.

EPE (2011). *Projeto da usina hidroelétrica de Belo Monte – fatos e dados*. Available at: www.mme.gov.br/documents/10584/1590364/BELO_MONTE_-_Fatos_e_Dados.pdf/94303fc2-d171-45be-a2d3-1029d7ae5aad. Acessed in: 09 October 2017.

Fainguelernt, M. (2016). The historical trajectory of the Belo Monte hydroelectric plant's environmental licensing process. *Ambiente and Sociedade*, V. XIX, n. 2, 245–264.

Fearnside, P. M. (2006). Dams in the Amazon: Belo Monte and Brazil's Hydroelectric Development of the Xingu River Basin. *Environmental Management*, 38(1), 16–27.

Fearnside, P. M. (2009). Brazil's environmental policies for the Amazon: Lessons learned from the last 20 years. Panel contribution on Models of Development: An Analysis of the Last 20 Years of Public Policies for the Amazon Region.

Fearnside, P. M. (2016). Brazil's Amazonian forest carbon: the key to Southern Amazonia's signifcance for global climate. *Regional Environmental Change*, 18(1): 47–61.

Fernandes, E. (1992). Law, politics and environmental protection in Brazil. *Journal of Environmental Law*, 4(1), 41–56.

Graeff, B. (2012). Should we adopt a specific regulation to protect people that are Displaced by Hydroelectric Projects?: Reflections based on Brazilian law and the Belo Monte Case. *Florida A and M University Law Review*, 7(2), 262–285.

Gramacho, M. (2010). Ibama concede licença ambiental prévia para a usina hidrelétrica de Belo Monte. Available at: www.mma.gov.br/informma/item/6071-ibama-concede-licenca-ambiental-previa-para-a-usina-hidreletrica-de-belo-monte. Accessed 9 October 2017.

Grove, C. (2014). To build a global movement to make human rights and social justice a reality for all. *Sur International Journal on Human Rights*, 11(20), 352–362.

Guatam, A., Haubold, I., Pacey, V., Papirnik, D., Premjee, M., Schlumpf, P., and Tolley, G. (2014). Brazil's Belo Monte: A cost-benefit analysis. Available at: http://franke.uchicago.edu/bigproblems/BPRO29000-2014/Team09-EnergyPolicyPaperBeloMonte.pdf. Accessed 1 April 2018.

Harari, I. (2017). Licença de Belo Monte é novamente suspensa. Available at: www.socioambiental.org/pt-br/noticias-socioambientais/licenca-de-belo-monte-e-novamente-suspensa. Acessed 9 October 2017.

International Energy Agency (IEA) (2006). Brazil (Partner country). The OECD. Available at: www.iea.org/countries/non-membercountries/brazil/ Accessed 1 April 2018.

IEA (2016a). Brazil energy indicators. The OECD. Available at: www.iea.org/statistics/statisticssearch/report/?year=2013andcountry=Brazilandproduct=RenewablesandWaste Accessed 1 April 2018.

IEA (2016b). Brazil country analysis. The OECD. Available at: www.eia.gov/beta/international/analysis_includes/countries_long/Brazil/brazil.pdf Accessed 1 April 2018.

International Rivers (2016). The Belo Monte Dam. *International Rivers*. Available at: www.internationalrivers.org/campaigns/belo-monte-dam, Accessed 1 April 2018.

ISA (2015). Dossiê – Belo Monte: não há condições para a licença de operação. Available at: www.socioambiental.org/pt-br/noticias-socioambientais/dossie-belo-monte-prevencao-e-mitigacao-falham-e-obra-provoca-novos-impactos-sobre-povos-indigenas. Acessed 9 October 2017.

Jaichand, V., and Sampaio, A. A. (2013). Dam and be damned: the adverse impacts of Belo Monte on in indigenous peoples Brazil. *Human Rights Quarterly*, 35(2), 408–447.

Luomi, M. (2014). *Sustainable Energy in Brazil: Reversing Past Achievements or Realizing Future Potential*. Oxford: Oxford Institute for Energy Studies.

Meisen, P. (2010). Renewable energy potential of Brazil. *The Global Energy Network Institute*. Available at: www.geni.org/globalenergy/research/renewable-energy-potential-of-brazil/re-pot-of-brazil.pdf Accessed 1 April 2018.

Moreira, J. M. L., Cesaretti, M. A., Carajilescov, P., and Maiorino, J. R. (2015). Sustainability deterioration of electricity generation in Brazil. *Energy Policy* 87: 334–346.

Norte Energia (2016). The Belo Monte Dam Complex. The Government oF Brazil. Available at: http://norteenergiasa.com.br/site/ingles/belo-monte/. Accessed 1 April 2018.

Oliver-Smith, A. (2014). Framing socio-environmental justice by Amazonian indigenous peoples: The Kayapo Case. *Tipiti: Journal of the Society for Anthropology of Lowland South America*, 12(2), 6.

Oxford Analytica (2014). *BRAZIL: Sustainable Energy Faces Rising Obstacles*. Oxford: Oxford University Press.

Petras, J. (2013). Brazil: Extractive capitalism and the great leap backward. *World Review of Political Economy*, 4(4), 469–483.

Pontes, N. (2017). Após um ano de funcionamento, Belo Monte segue envolta em polêmicas. In: Carta Capital, June 2017. Available at: www.cartacapital.com.br/sociedade/apos-um-ano-de-funcionamento-belo-monte-segue-envolta-em-polemicas. Accessed 9 October 2017.

Randell, H. (2015). Structure and agency in development-induced forced migration: the Case of Brazil's Belo Monte Dam. *Population and Environment*, 37(3), 265–287.

Reis, J., and Souza, J. (2016) Grandes projetos na Amazônia. A hidrelétrica de Belo Monte e seus efeitos na segurança pública. *DILEMAS – Revista de estudos de conflito e controle social*, 9(2), 215–230.

Steffek, J., and Romeiro, V. (2016). Private actors in transnational energy governance. In: M. Knodt and N. Piefer, eds, *Challenges of European External Energy Governance with Emerging Powers*. Farnham: Ashgate.

Swedish Agency for Growth Policy Analysis (2013). *Energy Policy in Brazil. Perspectives for the medium and long term*. Östersund: SAGPA.

Watts, J. (2016). Belo Monte dam operations delayed by Brazil court ruling on indigenous people. *Guardian*. Retrieved from www.theguardian.com/world/2016/jan/15/brazil-belos-monte-dam-delaycourt-indigenous-people

Westholm, L., Henders, M. O. S., and Mattsson, E. (2011). Learning from Norway – A review of lessons learned for REDD+ Donors. Focali Report. Gothenburg, Sweden:

Focali. Available at: http://forestbonds.eu/sites/default/files/userfiles/1file/Focali%2520 Report%2520No3%25202011_Norway%2520REDD.pdf, Accessed 1 April 2018.

World Bank (2016). Brazil development indicators. The World Bank Group. Available at: http://data.worldbank.org/data-catalog/world-development-indicators. Accessed 1 April 2018.

Zanotti, L. (2015). Water and life: hydroelectric development and indigenous pathways to justice in the Brazilian Amazon. *Politics, Groups, and Identities*, 3(4), 666–672.

Zortéa, M., Bastos, N., and Acioli, T. (2015). The bat fauna of the Kararaô and Kararaô Novo Caves in the area under the influence of the Belo Monte Hydroelectric Dam, in Pará, Brazil. *Brazilian Journal of Biology – Revista Brasleira De Biologia*, 75(3), 168–173.

6 An assessment of UN-REDD in Lam Dong Province, Vietnam

Laura Maxwell, Vidya Nair, Stephanie Soloman and Larry Swatuk

Introduction

The United Nations Programme on Reducing Emissions from Deforestation and Forest Degradation (referred to as UN-REDD and REDD+), is designed to provide financial value for carbon stored in forest, whereby forest managers and users aim to reduce emissions and invest in low-carbon options. With a focus on developing countries, the programme is intended to encourage policy approaches and positive incentives that promote conservation, sustainable management of forests, and enhancement of forest carbon stocks (Thompson *et al.*, 2011). Within the programme, countries including Vietnam will measure and monitor CO_2 emissions resulting from deforestation and degradation within their borders. Forests serve a number of important ecological functions, and deforestation and degradation are known to negatively contribute to soil erosion, flooding, loss of biodiversity, and impaired water supply and quality (Zaiha *et al.*, 2015). Blue water is associated with the flow of surface and groundwater (UNESCO, 2006), while green water encompasses 'precipitation on land that does not run off or recharge groundwater, but is stored in the soil or temporarily stays on top of the soil or vegetation' (Hoekstra *et al.*, 2011). Thus, initiatives including those delivered by the UN-REDD may work to reduce emissions, while protecting and/or improving environmental health and water security.

As a method to mitigate the effects of climate change, prevent land degradation and protect green and blue water, the Government of Vietnam launched its UNREDD national programme for forest protection and restoration in 2009. This chapter discusses challenges with government policy for land tenure and resource management, and the importance of the principle of free, prior and informed consent (FPIC) and appropriate payment for forest environmental services (PFES), which may act as a barrier to meaningful participation. If Vietnam's UN-REDD Programme is to achieve success, the government must revise its national priorities, revise its land allocation policy and ensure meaningful stakeholder participation. The chapter first discusses some of Vietnam's economic goals and policies as a means to give context to some of the drivers of deforestation that will be addressed. Following which, factors that could hinder the success of the UN-REDD programme in Vietnam are discussed; these issues surround the developmental policies that act as drivers of deforestation and issues surrounding the state of land tenure in Vietnam. Analysis of the importance of meaningful participation in the form of FPIC and system of PFES as integral parts of forest governance are also highlighted followed by discussion on the future of UN-REDD in Lam Dong, Vietnam.

Vietnam's goals for development, environmental sustainability and forests

Vietnam, a country in Southeast Asia, borders China, Cambodia and Laos, and is considered to be one of Southeast Asia's fastest growing economies. Furthermore, Vietnam has an ambitious goal to become fully developed by 2020 (BBC, 2016). Prior to its significant economic development in the mid 1980s Vietnam suffered from vast food insecurity due to poor central planning of farming and land management. Additionally, at this time there were many signals that the fall of the Soviet Union was soon approaching, indicating that Vietnam would soon lose financial support. Both of these factors induced a need for a change in economic practice (Kirk and Tuan, 2009). In 1986, the Central Government pushed for economic modernization under Doi Moi (Renovation) policy. The purpose of the Doi Moi policy was to initiate a change from a centrally planned economy to a socialist-oriented market economy to induce modernization of the country's infrastructure, and to open doors for external partnerships (Kirk and Tuan, 2009; Lan, 2011). Consequently, Vietnam has gone from being one of the poorest countries to one of exponential growth, much of this resulting from the success of the agricultural sector (Hiebert and Nguyen 2012; Kirk and Tuan, 2009).

Since Vietnam embraced market liberalization, the United States has become one of the country's largest trading partners (World Bank, 2015). In 2007, Vietnam joined the World Trade Organization, and by 2010 the nation joined the Trans-Pacific Partnership for Free Trade. In its progressive push for development the Central Government, in partnership with the World Bank, developed the Socio-Economic Development Strategy (SEDS) from 2011–2020. SEDS is a strategy for reforms to Vietnam's economic structure, increasing its environmental sustainability, social equity and macro-level stability. Within the first five years, the programme improved living standards of the poor and produced economic growth. However, the World Bank (2015) also notes there has been slow progress towards the goals initiated by SEDS as of late. Nonetheless, even with the developmental progress made, Vietnam's economy remains reliant on the forest and agricultural sectors (approximately 60 per cent of the workforce in Vietnam is agriculture-based), although as a proportion of GDP the sector's labour contributions are continually being reduced in favour of light and heavy industries, as important components of the country's economic goals (BBC, 2016; Hiebert and Nguyen, 2012; Lan, 2011).

A high emphasis on cash crops and logging, both of which contribute to soil degradation, have been and still are major contributors to Vietnam's GDP;

consequently this past activity has necessitated a need for reforestation (UN-REDD, 2009; McElwee, 2004). Vietnam's forest cover progressively declined from the years of 1943 to 1990 (43 per cent forest cover to 27 per cent) as a result of plantation of non-indigenous trees, and logging for export by State Forestry Enterprises (SFEs) (UN-REDD, 2013; McElwee, 2004). These issues have made investment in fertile land a national priority (World Bank, 2015; UN-REDD, 2009). To mitigate these environmental issues, Vietnam launched a UN-REDD pilot project in Lam Dong Province in 2009, with a strong focus on the conservation of biodiversity and prevention of land degradation (UN-REDD, 2013).

UN-REDD in Lam Dong Province

Lam Dong Province borders the provinces of Khanh Hoa, Ninh Thuan, Dong Nai, Binh Thuan, Binh Phuoc, Dak Lak, Dak Nong, and has 61 per cent forest coverage. Currently, Lam Dong is considered by the central government as a key economic zone in the south. This is due to the province's high potential for economic growth of which industrial bushes, forestry, minerals, tourism and animal husbandry are considered key areas for economic development – especially in light of Vietnam's 2020 SEDS goals (World Bank, 2015; Phương, 2011). In 2009, Lam Dong Province was chosen as a pilot site for the UN-REDD PFES conservation and biodiversity programme (see Table 6.1 for specific locations of where the pilot took place) (UN-REDD, 2013). Chiramba *et al.* (2011) note that the initial project incurred a number of successes regarding its public awareness campaign, payment scheme, and the inclusion of local groups into the formulation of the final project. In addition to this the initial rainforest valuation study gave the central government confidence in the implementation of PFES for farmers in the community. Furthermore, as a result, farmers were paid four times as much under PFES than under previous national conservation schemes. These payments encouraged more sustainable forest land management practices among farmers in the province (Enright, 2013; Chiramba *et al.*, 2011). The project provided important scientific evidence that intact forests would reduce soil erosion and future costs for hydro- and water-supply companies in the province. Upon its successful completion in Lam Dong, the project was expanded into other parts of the country, and has since been deemed a centre of excellence. Moreover, places like Cambodia have looked at the project as a framework for similar projects in their nations. However, with the successes there were also challenges in the initial project regarding: mitigating competing values, insufficient funding for conservation efforts, collaboration between ministries, and issues surrounding land law and use rights (Chiramba *et al.*, 2011).

The state of land tenure in Vietnam

While the UN-REDD programme was being expanded into the rest of Vietnam in 2012 land issues came to the forefront of political activity at the same period. Seventy per cent of complaints filed to the government that year were regarding land, due to disputes between villagers, the government and developers; tensions arose in cases where villagers were not consulted in decisions to change land use for re-development and construction (Hiebert and Nguyen, 2012; USAID, 2012). In some cases these disputes resulted in relocation without consultation. Hiebert and Nguyen (2012) propose that the political significance of land law, specifically regarding use rights, is due to the economic importance of the agricultural sector in Vietnam. As previously mentioned, farmers comprise 60 per cent of the country's workforce. A large proportion of these farmers' 20-year land certificates and use rights were coming to an end in 2012. Traditionally, after the 20 year period land would then be re-allocated by the state to make room for new farmers. In 2012, there was speculation that the newly proposed land law for 2013 would allow for some private ownership, although this suggestion ultimately did not come to fruition. Furthermore, Hiebert and Nguyen (2012) suggest that some of Vietnam's political elite wanted to use the proposed 2013 land-law reforms as a means for larger political reform, which would lead to empowerment of the people; however, to the Central Government land-law reforms simply work to assert its power and establish its legitimacy. During this period of reforms, the United States considered lessening its ties to Vietnam if the Central Government did not improve its human rights record. Some of these human rights issues directly relate to treatment of farmers, and villagers in land disputes (Hiebert and Nguyen, 2012). Under the 2013 land law, the state still manages and allocates land for the people of Vietnam; however, restrictions have been placed on the government's ability to obtain land for redevelopment (ITPC, 2016. Furthermore, land certificates in farming areas are now issued for 50 years as opposed to 20 years (ITPC, 2016).

In Vietnam, the land is owned by the people, but overseen and managed by the state, which administers use-rights to individuals, households and communities based on family size and financial ability to work the land (Larson *et al.*, 2013; USAID, 2012; Kirk and Tuan, 2009). However, individuals, households and communities do not own the land outright – this is called usufruct rights. Land-use rights, or land tenure, are rights that allow individuals to use the land they live on. However, the state has the right to intervene and dictate how land is used for those who hold land-use rights certificates (LURC) based on yearly government quotas for food security, and Central Governmental development goals (ITPC, 2016; USAID, 2012; Kirk and Tuan, 2009). Under law, LURCs can be inherited, rented or sold (see Table 6.1 for a breakdown of successive land laws). Issues raised by LURC holders and communal users (these groups do not have LURCs, but manage land communally) are: lack of government transparency, the duration of time LURC holders have between contract renewals, government delays in the decision-making process to grant tenure,

Table 6.1 Changes in use rights and land law

Land law year	Agricultural Land-Use rights			
	Possessors	Tenure	Maximum holdings allowed	Subrights
1988	Generally, provision was made for the state to allocate land for use by state farms, cooperatives, production groups, state enterprises, army units, state institutions, social organizations and individuals	This Law did not regulate, but the practice was 5–15 years	No specific regulation but Law gave authorities the right to decide depending on the local situation of land areas and peasant numbers	None; this Law even forbade the purchase, sale or lease of land use rights for money under any circumstances
1993	3 groups: (1) state agencies; (2) households and (3) individuals	20 years for land cultivated with annual crops and 50 years for land cultivated with perennial crops	3 ha, but sub-provisions placed a limit of 2 ha/possessor in the north and 3 ha/possessor in the south	5 subrights, namely, the right to transfer, exchange, lease, inherit and mortgage land-use rights
1998	No major changes to former version of Land Law			
2001	No major changes to former version of Land Law			
2003	7 groups: (1) domestic organisations (2) domestic family households and individuals (3) residential communities [domestic] (4) [domestic] religious institutions (5) foreign organisations with diplomatic functions (6) overseas Vietnamese (7) foreign organisations and individuals with investments in Vietnam		3 ha/possessor for land cultivated with annual crops; 10 ha/possessor for land cultivated with perennial crops	10 sub-rights: the right to (1) sub-lease land use (2) grant land-use rights (3) secure land-use rights as a collateral (4) employ land-use rights as a form of capital (5) claim compensation if land-use rights are seized by the state (6–10) the 5 sub-rights stated in 1993 Land Law

Source: author's synthesis from the 1988 Land Law and amended versions of Land Law passed in 1993, 1998, 2001 and 2003.

payment and allocation of land type. In addition, there have been conflicting demands for use rights of forested areas between farmers and forest users, most of which are vulnerable populations and indigenous groups. Furthermore, most minorities and indigenous groups that rely on the forest do not have formally recognized forest use rights and are not considered official land forest tenants; only 4 per cent of these populations reliant on the forest have secure land tenure (Do *et al.*, 2012; USAID, 2012). As the forest is managed by state minorities and indigenous groups, they would like greater recognition and involvement in forest management through increased use rights – specifically in protected forest areas (Do *et al.*, 2012; USAID, 2012).

Forests in Vietnam are divided into three classifications, according to type and use. The first is *special use forests* (approximately 15 per cent of forested area in Vietnam), encompassing national parks and conservation sites managed directly by the state, where exploitive activities are prohibited (USAID, 2012; McElwee, 2004). The second classification is *protection forests* (approximately 38 per cent of forested area); 30 per cent of this section is allocated for private sector use, which have limited production use and mitigate against soil erosion and desertification (USAID, 2012). These forests are primarily managed by the People's Committee (CPC), which receives its direction from the Central Government (McElwee, 2004). Finally, *production forests* (approximately 46 per cent of forested area), 70 per cent of which is managed by the private sector for commercial use, managed by SFEs, individuals, households or co-operatives (USAID, 2012; McElwee, 2004). Challenges encountered by UN-REDD Vietnam regarding land tenure and various other government policies and activities may have unintended consequences and must be dealt with if the programme is to achieve success.

Challenges with government policy for sustainable resource management

Policy to address drivers of deforestation in Vietnam

While the UN-REDD Programme in Vietnam has thus far been regarded as a success, the National REDD+ Action Plan (NRAP) of Vietnam does not have a specific strategy to directly address the drivers of deforestation and degradation, as recommended by the UNFCCC and the World Bank (Hoan and Catacutan, 2014). Drivers of deforestation in Vietnam include a combination of economic factors, institutions and national policies. The primary economic drivers of deforestation include timber and coffee production, shrimp farming in mangrove forest and rubber production in several provinces (Thuy *et al.*, 2012). The UN-REDD Programme has been criticized for several of its decisions regarding carbon, such as the inclusion of large tree plantation and industrial forest operations, which while a source of carbon sequestration could limit emissions reductions, and reduce coherency across REDD+ programmes (McDermott, 2014). However, it is not yet clear whether plantation forests will be included in REDD+ activities. Although the Government of Vietnam has identified the need to develop strategies to improve the sustainability of production in these sectors, in addition to mitigating emissions from deforestation and forest degradation, specific strategies have not been widely implemented (UN-REDD, 2013). In its Intended Nationally Determined Contribution (INDC) to the United Nations Framework on Climate Change (UNFCCC), the Government of Vietnam committed to several contributions, including a reduction of emissions intensity per unit of gross domestic product of 20 per cent compared to 2010 levels and an increase in forest cover of 45 per cent by the year 2030 (INDC VN, 2015). The REDD+ programme in Vietnam will therefore play an important role in reaching the commitments described in its INDC for emissions reductions targets, sustainable forest management, afforestation and reforestation, and livelihood development. The omission of policy to address the underlying factors of deforestation in its REDD+ programme impede delivery of its INDC; and furthermore may indicate that the interests of Vietnam currently differ from the core objectives of the UN-REDD Programme.

One example of a major policy that contradicts the objectives of REDD+ is related to rubber-plantation expansion, with the goal of exploiting and intensifying the advantages of land resources and national conditions for sustainable development. Available data shows that from 1975 to 2008, land under rubber increased from less than 100,000 ha to more than 600,000 ha resulting in an increase in production from less than 50,000 tons to nearly 700,000 tons (UN-REDD, 2013). Government policy intends to continue expansion ultimately resulting in a stable area of 800,000 ha of rubber plantations in Vietnam by 2020 (Phuc and Nghi, 2014). The process of converting natural forest to rubber plantation is known to threaten biodiversity and ecological functions (Ahrends *et al.*, 2015). Since rubber plantation expansion has been identified by the Government of Vietnam as a source of economic growth, its strategy for economic growth contradicts the objectives of its UN-REDD programming. It is therefore important to identify areas in which government policy can be improved to reconcile differences between these two programmes. Forest classification systems and approval procedures in Vietnam for the conversion of natural forest for rubber expansion, or the establishment of tea and coffee plantations in Lam Dong Province, may endanger health and/or naturally regenerating forest area. Government criteria for degraded forest classification could result in the conversion of natural forest for economic productivity; recommendations for Phase II of UN-REDD in Vietnam calls for a review of current policy and criteria to approve rubber plantation establishment exclusively on bare land or poor agricultural land, and to augment current criteria for 'poor' forest to below $100 \text{ m}^3/\text{ha}$ and promote rubber planting within natural forested areas (UN-REDD, 2013). The plans to address drivers of deforestation as outlined in the UN-REDD Vietnam Phase II Programme, if successfully implemented, will work to address the underlying issues causing deforestation in Vietnam, enhancing the sustainability of the programme. However, if strategies are not implemented or properly enforced, the REDD+ initiatives in Vietnam will likely not be effective in reducing emissions from deforestation and forest degradation. As long as large-scale forest conversion continues and the drivers of such deforestation and forest degradation are ignored, the success of REDD+ in Vietnam remains uncertain.

Challenges with land tenure encountered by UN-REDD Vietnam

Prior to 1986, Vietnam experienced slow or stagnant economic growth and famine, and the forests and agricultural lands were controlled and operated by the state (Kirk and Tuan, 2009; McElwee, 2004). Because the state was sole proprietor of the forest via SFEs, people's communal forest practices to obtain food at this time were often deemed illegal by the government (McElwee, 2004). During the Doi Moi (Renovation) policy period after 1986, forest land reforms began to take place, creating a shift towards more individual and communal management (Do et al., 2012; Kirk and Tuan, 2009), according to current forest classifications. The Doi Moi policy established enhanced tenure security by establishing greater gender equity in the attainment of land tenure and liberalization of agricultural lands. These initiatives allowed for land sales in rural areas, which had created further prosperity through food security and higher living standards. Furthermore, under this policy Vietnam became the world's third largest exporter of rice, and agriculture became a driver of economic growth (Kirk and Tuan, 2009). Nevertheless, under the current system, SFEs and other stakeholders, such as developers, still have greater access to use rights and management capability in comparison to households, individuals and communities (Do et al., 2012). Larson et al. (2013: 680) attribute this is to,

> 1) political and economic interests of actors competing for forest land and resources; 2) limited technical, human and economic capacity to carry out accurate and effective demarcation and titling; and 3) ideological barriers, such as opposition to or concerns with the idea that forest dwellers can be effective forest stewards.

In addition, the use rights of minorities and indigenous groups are often limited to production forests, which, more often than not, have poorer quality soil, henceforth, limiting their ability to work the land properly (Larson et al., 2013).

Kirk and Tuan (2009), point out even more challenges to tenure for minorities and indigenous groups. These include challenges to consolidate reforms due to increasing disparities in income between regions and urban and rural areas; misalignment of power in allocating land, and landlessness from emergency land sales (resulting when people are unable to pay for the land). These sales disproportionately affect minority and indigenous groups, because they are the least likely to have consistent sources of credit. For Kirk and Tuan (2009: 2),

> land-tenure reforms have not yet increased rural people's [primarily minorities and indigenous groups] access to credit.... One reason is that formal

banking institutions still seem reluctant to accept land-use certificates as collateral, believing that the land will be hard to seize in case of credit default.

The impact of tenure legislation on potential success of UN-REDD Programmes is observed in several countries. Do *et al.* (2012) and Cotula and Mayers (2009) suggest that although current tenure structure may not block the purposes of REDD+, a re-arrangement of the tenure system and the expedition of allocation would make the delivery of REDD+ initiatives more successful. Cotula and Mayers (2009) make a distinction between having secure tenure rights and insecure rights. Secure tenure acts as a leverage point for minorities and indigenous groups in their interactions with government regarding forest management. In this case, Land Certificates have been given to a smallpercentage of forest dwellers in Vietnam. However, many state reforms to land law have had little influence on the ground for minorities and indigenous groups (USAID, 2012). This is because community understanding of use rights differs from the state's definitions. Furthermore, these groups are often poorly informed of their use rights, thereby causing their underutilization. Further still, obtaining use rights is dependent on an individual's ability to obtain financial and human capital, both of which are lacking for many minorities and indigenous peoples. These factors thereby limit indigenous groups' and minorities' ability to hold land-use certificates, limiting their full participation in REDD+ projects (Larson *et al.*, 2013).

Insecure land tenure increases vulnerability to dispossession, encroachment, land grabbing and illegal land sale to stakeholders wanting to make money from PFES (Larson *et al.*, 2013; USAID, 2012; Cotula and Mayers, 2009). As mentioned previously, emergency land loss in Vietnam is disproportionately experienced by the rural poor consisting of minority groups and indigenous peoples (Kirk and Tuan, 2009). Another form of insecure land tenure is in the case of unallocated land in Vietnam. Unallocated land is not a formal classification under the three types of forest spoken of previously, so it is not legally recognized in order for its users to obtain tenure. Forest use rights are given primarily in production forest, often with poorer quality soil (USAID, 2012), and SFEs typically own best forest areas. This therefore puts government stakeholders and other powerful actors into a position where they could gain more from REDD+ than the community, thereby limiting minority and indigenous activity in forest management, and these forest managers receive little compensation for their effort (Larson *et al.*, 2013; Do *et al.*, 2012).

Challenges to meaningful participation in the UN-REDD Vietnam programme

The importance of free, prior and informed consent

Meaningful participation is a key instrument for good forest governance, and has been identified as an important component to the success of UN-REDD

Programmes. Several studies have found that conflict over forests is often the result of feelings of injustice due to restricted access rights to forest resources. These feelings are often due to a lack of information communicated to local communities on REDD+ programme implementation, leading to poor understanding and reduced participation in projects (Patel *et al.*, 2013). Conflict between and among government, state-owned enterprises and local communities has occurred in several sectors in Vietnam, particularly over forest management. SFEs in Vietnam are responsible for the management of natural and planted forests for timber production. Recently, conflict has been escalating between SFEs, local farmers and communities competing for control over land, as SFEs have restricted local forest access and are competing for use of resources. In Cot Coi, 25 per cent of community members had no access to forested areas and limited available productive land for cultivation due to government agency regulations; villagers are currently cultivating acacia trees on approximately 42 per cent of government agency land, resulting in quarrels, destruction of seedling trees and exploitation of community profits (To *et al.*, 2015). Ongoing and escalating conflict between government agencies and communities over forest resources has, in some areas of Vietnam, reduced the legitimacy of government authorities, as SFEs are the first point of government contact for many Vietnamese, especially in rural areas (Sikor and Cầm, 2016). Although SFEs are not operating directly under Vietnam's UN-REDD Programme, conflict over forested land and resources could very likely appear in government REDD+ projects if government officials do not ensure that information is clearly communicated to community leaders and members. The government should also promote enhanced and meaningful participation that is non-exploitative in order to reduce the potential for further conflict. If the Government of Vietnam is not able or willing to engage stakeholders in its programming in a meaningful, and equitable way, its potential for success in establishing an effective conservation and reforestation programme will be greatly reduced (Sikor and Cầm, 2016).

In its evaluation and verification of FPIC, an external auditor found that information provided to local peoples by the UN-REDD Vietnam Programme was not sufficient to meet the 'informed principle' of FPIC. This was due to time constraints during community meetings, with focus placed solely on the potential benefits of the programme, giving little to no attention to potential risks and costs related to REDD+ projects for local peoples (Quang *et al.*, 2010). A lack of participation of indigenous people in the National REDD Network (NRN) has also been identified. This could be due to negative perceptions of ethnic minorities by the Vietnamese Government and conservation authorities, giving preferential treatment to ethnic minorities when it comes to social services, but excluding them from decision-making (Hoan and Catacutan, 2014). However, despite these issues, in a study of the acquisition of FPIC in 16 of the 41 UN-REDD projects in Vietnam, consultation and participation were found to be well above minimum levels, with no observable differences in regional participation; projects restricting land use demonstrated lower levels of participation (Lawlor *et al.*, 2013). As demonstrated in the case of Cot Coi, as well as across much of the Globa South, if communities are not informed and included in decision-making, conflict can arise and escalate over time (Jeffery and Vira, 2001). The Government of Vietnam must therefore ensure that the principles of FPIC are met. If FPIC is not completed for UN-REDD projects in Vietnam, there will likely be an increased risk of conflict between government and communities. A lack of meaningful participation in the UN-REDD Vietnam projects may result in a boomerang effect, as communities have demonstrated worldwide their willingness to fight for the forest (Jeffery and Vira, 2001). Diminishing the potential success of the programme will be the least of the state's problems in the face of persistent pockets of localized violence. In addition to enhancing community participation through the acquisition of FPIC prior to the implementation of projects, an important and challenging component of participation in REDD+ projects is payment for ecosystem services.

Ensuring appropriate and equitable payment for ecosystem services

The Government of Vietnam is currently in the process of developing a comprehensive benefit-distribution mechanism to deliver revenue to participants of the UNREDD programme in a transparent, equitable and cost-effective manner. A 'pro-poor' system for the payment for ecosystem services (PES) is designed to encourage economic and/or social development. The widely accepted principles for design of a 'pro-poor' PES approach are the inclusion of stakeholders in decision-making on benefits and their distribution, the promotion of cost-effective distribution, efficiency and transparency, the promotion of livelihoods, and the equitable distribution of benefits (Ogonowski and Enright, 2013). The PFES system in Vietnam attempts to follow these principles, in addition to applying a gender-inclusive approach, but has encountered several challenges. Centralized state forest land management in past government reforestation projects in Lam Dong province has in certain cases been found to constrain local households from receiving PFES benefits, triggering conflict in villages (To *et al.*, 2012; cf. Bayrak and Marafa, 2017). The importance of identifying ways of structuring payments to avoid negative outcomes is also evident based on findings that explicit use of cash incentives has frequently resulted in participants dropping out of the programme. In some cases, payments offered for conserving or restoring natural forest area have been lower than what participants could earn harvesting high-value crops, leading to unsuccessful land-use conversion in the area (Hoang *et al.*, 2013). Strategies to avoid these phenomena may include bundling PFES with support for sustainable agroforestry, allowing locals to supplement their income. The use of non-cash incentives, such as clear land tenure and technical assistance, is another alternative, which could provide appropriate and cost-effective forms of payment for remote and indigenous populations (Hoang *et al.*, 2013). Since monetary incentives cannot currently act as a substitute for income earned through land cultivation, the use of non-cash incentives could supplement lost income in favour of forest protection. Alternative strategies such as these could therefore make participation in REDD+ projects more attractive and beneficial to small land holders and marginalized populations (Huynh and Keenan, 2017).

Disparities in income level and land allocation have been widening in Vietnam, where poor households are more likely to have small areas of cultivable land as a result of governance decisions (Hiebert and Nguyen, 2012). Generally, REDD+ benefits are distributed on the basis of landholding size, which means that large landholders are awarded disproportionately greater benefits than are the poor. To create an equitable distribution system, the Government of Vietnam would need to ensure that weak claimants are not excluded or marginalized by the benefit-distribution system (see Wood et al., 2018 for interesting ideas about achieving social justice through the clean development mechanism (CDM)). In order to distribute the costs and benefits of participation in REDD+ schemes in a more equitable manner, it has been suggested that the benefit-distribution system could mandate declining rewards for each additional unit of land, encouraging participation of small landholders (Schwarte and Mohammed, 2011). Ecuador's Socio Bosque programme was based on this method, reducing payment per area as land enrollment increased, with the majority of payments being made in the form of cash or technical assistance; despite the system's failure to distribute contracts efficiently, 50 per cent of households surveyed reported increased household income, and alternative forms of payment distribution could improve outcomes in future trials (Hejnowicz et al., 2014).

The failures and successes of Ecuador's attempt at a descending payment scale for PFES demonstrate that there is currently no one 'best option' that could be employed in Lam Dong. The Provincial Governments of Vietnam have already, in some cases, made progress towards the pro-poor objectives of the national PFES system. For example, in 2009, forestry and government officials in Lam Dong province agreed to allocate 10 per cent of PFES payments to cover provincial expenses, 9 per cent to large forest owners, and 81 per cent to household payments (US$15/ha) (McElwee, 2012). Allocating a greater amount of PFES payments to the household level should theoretically work to ensure that UN-REDD payments reach those they are intended for, including small land holders, marginalized people and indigenous populations. The Government of Vietnam has made progress, but must develop structures of forest governance that are aligned with the objectives of REDD+ and encourage active participation from local communities regardless of gender or ethnicity, forest restoration and protection programmes in Vietnam. Failure to do so may result in greater numbers of participants deciding not to participate in PFES schemes or abandoning the programme altogether if payments are not appropriate or useful to them. Failure to distribute PES equitably may escalate pre-existing tensions and conflict between the government and communities (Huynh and Keenan, 2017).

The future of UN-REDD in Lam Dong, Vietnam

While the UN-REDD scheme in Vietnam continues to operate, in some ways quite successfully, the true future of the venture remains uncertain (Huynh and Keenan, 2017; Sikor and C m, 2016). As previously mentioned, themes of meaningful participation through FPIC along with PFES continues to be a challenge. Moreover, forest and land reforestation policy along with strategies for land tenure continue to be highly contentious and even more difficult to formally analyze. Thus, while the mechanisms in place are currently in the process of developing Lam Dong's environmental sustainability and enhancing forest conservation within the country, the UN-REDD Programme poses a number of challenges to consider in addressing this complexity.

Trade-offs

Expanding the current carbon capture storage within Vietnam's forests is beneficial towards the goal of addressing blue- and green-water challenges facing much of the developing world (Sunderlin *et al.*, 2013). Expanding the REDD+ concept towards phase two within Lam Dong Province is also beneficial, as it continues to engage with indigenous and rural communities who are viewed and accepted as experts in the field of forest management. However, the nexus of all these exchanges is the idea of trade-offs. The balance of a successful UN-REDD programme is achieved between multiple desirable yet incompatible features. These trade-offs require more of the Government of Vietnam's focus as well as attention from UN-REDD to ensure that the future of Lam Dong and Vietnam continue towards a positive environmental trajectory (Suhardiman *et al.*, 2013).

While showcasing plans to secure further opportunities for small-scale stakeholders, such as indigenous communities, the opportunity to be forest managers and store carbon credits may not be as lucrative as once thought (Locatelli *et al.*, 2011). Market prices for PFES are sensitive, as it is foreign entities that offer prices that have been determined internationally (IUCN, 2010). For example, the opportunity cost of rubber is US$95/tonne and remains outside the expected range of the payment for the ecosystem-services scheme. With the indigenous communities understanding this lucrative market and the Government of Vietnam continuing to invest in rubber plantations, the collaborative model between the two stakeholders could hinder the success of a UN-REDD venture (Phuc and Nghi, 2014). This means, rather than communities being consulted and engaged as to appropriate payments for forest management, transactions are simply being carried out while cultural appropriateness is excluded from programme introduction. Ogonowski and Enright (2013) advocate for a pro-poor model to be at the centre of REDD+ designs, whereby supporting mechanisms are in place to pursue both economic and environmental development for marginalized populations. Thus, the trade-off of an internationally set market price for carbon may not be the most equitable solution, as cultural sensitivity and local engagement should be increasingly valued (Wood *et al.*, 2018).

Moreover, with the REDD+ scheme relying on payment for ecosystem services, challenges arise due to government corruption and weak legal-regulatory frameworks to ensure equitable remittances. Barr and Sayer (2012) analyze the political economy of reforestation and forest restoration in Asia-Pacific, deeming it to be one that benefits states and political elites. Thuy et al. (2014) observe that the Government of Vietnam is typically directly involved with the payment scheme acting as an intermediary between buyers and sellers of carbon. With this, the government issued Decision 380, which piloted payments for environmental services within pre-existing government frameworks and also being the first country in Asia to undertake this national scheme (Thuy et al., 2014). While Decision 380 attempts to speak towards good governance, many other challenges came to fruition, including very low payments for intended populations, which do not have any effect on poverty reduction. This is due to the Government of Vietnam continuing to 'facilitate corruption and financial fraud, in some cases on a grand scale' and turn a blind eye to financial good governance (Barr and Sayer, 2012: 16). While Decision 380 is a step towards comprehensive forest development policy to create conditions of inclusivity within pre-existing resources, it fails to further the country's economic development due to increasing levels of corruption and misconduct.

Continued emphasis on mitigation efforts

Climate change mitigation is aligned with REDD+ proposals to recognize the rights of indigenous populations, respect the principles of FPIC along with securing progressive policy for sustainable forestry management (Griffiths and Martone, 2009). However, with this are efforts that seek to fulfill long-term goals, while also working to protect communities and strengthen the resilience of Vietnam's economy to adequately meet the current needs of the people today.

Mitigation, as defined by IPCC (2001: 8) seeks to 'reduce the sources or enhance the sinks of greenhouse gases [while] adaptation is an adjustment in natural systems in response to climatic effects which moderates harm'. According to a report from USAID (2011), Vietnam's Ministry of Agriculture and Rural Development works towards climate change by efficiently managing natural resources, such as forestlands and special use forests. These efforts encompass more of a mitigation tactic by reducing emissions and also attempt to build resilience of the indigenous peoples through the REDD+ network (Thuy et al., 2014).

Opportunities for meaningful exchange

In order for all stakeholders to benefit and for indigenous communal rights to be protected, a true collaborative approach must be adopted. This will also guarantee environmental protection by ensuring forest management needs are met and blue-water security is achieved. By reducing emissions from deforestation and forest degradation, an opportunity for meaningful exchange presents itself for all positions to be respected.

The UN-REDD offerings encourage a green-water approach by tackling various ecosystems such as community-managed forests, national parks, reforestation and afforestation efforts (Kenney *et al.*, 2015). This means that programme focus continues to be on carbon capturing to meet the country's biodiversity goals. However, here lies an opportunity for further meaningful exchange by not restricting the REDD+ scheme to only forests, but to blue-water pathways as well. RECOFTC (2007) note that within Lam Dong province, approximately 3.3 million ha of forests are held under local tenure, and 2.6 million ha under temporary management of communal authorities. Thuy *et al.* (2012) further observe that the forests within Lam Dong are natural forest, primarily mixed hardwood; within these, indigenous species are planted such as acacia, pine and bamboo. By ensuring local forests are intact, natural cycles of precipitation and land use are aligned with natural surface water and groundwater pathways (Berkes and Folke, 1998). Thus, conservation efforts remain a priority, as large portions of Vietnam's total forest area require proper management. Moreover the REDD+ design requires a strategy to address both blue- and green-water stresses.

Additionally, with entrusting local actors within Lam Dong to manage and care for the forests, the UN-REDD Vietnam Phase II Programme continues to ensure that stakeholder engagement is a priority. Pratihast *et al.* (2013) explain that by entrusting local populations with forest management, it not only protects the environment but also assures economic benefit through the PFES scheme. Furthermore, indigenous rights are valued and remain intact. Barr and Sayer (2012) reflect that often times participating villagers have very little leverage in negotiating agreements within conservation efforts, and while there are several ways to ensure meaningful participation, an opportunity lies within the UN-REDD programme within Vietnam to further develop these mechanisms and ensure accountability by having the programme operate in the interests of locals as well.

Conclusion

The UN-REDD Programme launched in Lam Dong Province, Vietnam is comprehensive in addressing current climate change issues of adaptation and mitigation respectively, and seeks to partner well with the national government. However, much like all programmes, it is far from perfect. While employing principles of FPIC and striving to allocate PFES, there continues to be a lack of transparency with indigenous populations, and minority groups as it relates to meaningful participation. Moreover, within the province the poor benefit-distribution mechanism plan presents a number of equity challenges. For Wood *et al.* (2018), this well-meaning programme often adds a socially problematic layer of complexity atop already existing strained social relations where, for example, forests are being fought over by community groups, private companies and state actors (see also Sikor and C m, 2016). The goals of the REDD+ scheme to ensure carbon sequestration and promote biodiversity within the region also compete with the Vietnamese government's economic and developmental goals for 2020 and its plans to expand rubber plantations within the country, thereby hindering the programme's objectives of forest protection and restoration. Vietnam's government policy as it relates to indigenous and minority communities' constrained ability to obtain LURCs and proper land allocation is also problematic, limiting these groups' ability to fully participate in REDD+. Thus, Vietnam's UN-REDD Programme goals to protect both green and blue water, while striving to achieve harmony within ecosystems, are complex, leaving its success contingent on both participation and policy reform.

References

Ahrends, A., Hollingsworth, P. M., Ziegler, A. D., Fox, J. M., Chen, H., Su, Y., and Xu, J. (2015). Current trends of rubber plantation expansion may threaten biodiversity and livelihoods. *Global Environmental Change*, 34, 48–58.

Barr, C. M. and Sayer, J. A. (2012). The political economy of reforestation and forest restoration in Asia-Pacific: Critical issues for REDD+. *Biological Conservation*, 154, 9–19.

Bayrak, M. M. and Marafa, L. M. (2017). Livelihood implications and perceptions of large scale investment in natural resources for conservation and carbon sequestration: Empirical evidence from REDD+ in Vietnam. *Sustainability*, 19(1802), 1–23.

BBC News (2016). BBC News-Vietnam country profile. Retrieved from www.bbc.com/news/world-asia-pacific-16567315

Berkes, F., and Folke, C. (1998). Linking social and ecological systems for resilience and sustainability. *Linking Social and Ecological Systems: Management Practices and social Mechanisms for Building Resilience*, 1, 13–20.

Chiramba, T., Mogoi, S., Martinez, I., and Jones, T. (2011). Payment for forest ecosystem services (PFES): pilot implementation in Lam Dong Province, Vietnam. Available at: fromwww.un.org/waterforlifedecade/green_economy_2011. Accessed 6 March 2016.

Cotula, L. and Mayers, J. (2009). Tenure in REDD – Start-point or afterthought? *Natural Resource Issues No. 15*. London: International Institute for Environment and Development (IIED).

Do, T. H., Catacutan, D., Vu, T. H., Lai, T. Q. (2012). Will current forest land tenure impede REDD+ efforts in Vietnam? *ASB Policy Brief No. 27*. Nairobi: ASB Partnership for the Tropical Forest Margins.

Griffiths, T., and Martone, F. (2009). *Seeing 'REDD'?: Forests, Climate Change Mitigation and the Rights of Indigenous Peoples and Local Communities*. Moreton-in-Marsh, England: Forest Peoples Programme.

Hejnowicz, A. P., Raffaelli, D. G., Murray, A., Rudd, P., and White, C. L. (2014). Evaluating the outcomes of payments for ecosystem services programmes using a capital asset framework. *Ecosystem Services*, 9, 83–97.

Hiebert, M., and Nuygen, P. (2012). Land disputes stir political debate in Vietnam. Center for Strategic and International Studies. Available at: http://csis.org/publication/land-disputes-stir-political-debate-vietnam. Accessed 27 March 2018.

Hoan, D. T. and Catacutan, D. (2014). *Beyond Reforestation: An Assessment of Vietnam's REDD+ Readiness*. Bogor, West Java, Indonesia: World Agroforestry Centre (ICRAF) Southeast Asia Regional Program.

Hoang, M. H., Do, T. H., Pham, M. T., Noodwijk, M., and Minang, P. A. (2013). Benefit distribution across scales to reduce emissions from deforestation and forest degradation (REDD+) in Vietnam. *Land Use Policy*, 31, 48–60.

Hoekstra, A. Y., Chapagain, A. K., Aldaya, M. M., and Mekonnen, M. M. (2011). *The Water Footprint Assessment Manual: Setting the Global Standard*. London: Earthscan.

Huynh, T. B., and Keenan, R. J. (2017). Revitalizing REDD+ policy processes in Vietnam: The roles of state and non-state actors. *Forests*, 8(53), 1–17.

Intended Nationally Determined Contribution of Viet Nam (INDC VN) (2015). Available at: www4.unfccc.int/submissions/INDC/Published%20Documents/Viet%20Nam/1/VIETNAM'S%20INDC.pdf. Accessed 27 March 2018.

IPCC (2001). *Climate Change 2001: Synthesis Report*. Cambridge: Cambridge University Press.

ITPC (Online Trade and Investment Information Portal) 2016. Government of Vietnam Land Law No. 45/2013/QH13. (2014). Available at: www.itpc.gov.vn/investors/how_to_invest/law/Law_on_land/mldocument_view/?set_language=en. Accessed 11 April 2016.

IUCN (2010). REDD in Vietnam: issues, opportunities, and linkages. Available at: www.iucn.org/about/work/programmes/forest/?6434/REDD-In-Vietnam-Issues-Opportunities-and-Linkages. Accessed 27 March 2018.

Jeffery, R. and Vira, B., eds (2001). *Conflict and Cooperation in Participatory Natural Resource management*. Basingstoke: Palgrave.

Kenney, L., Arvai, J., Vardhan, M., and Catacutan, D. (2015). Bringing stakeholder values into climate risk management programs: Decision aiding for REDD in Vietnam. *Society and Natural Resources*, 28(3), 261–279.

Kirk, M., and Tuan, N. (2009). Land-tenure policy reforms. Washington, D.C: International Food Policy Research Institute (IFPRI).

Lan, T. (2011). Overview of Vietnam's economy. Available at: www.vietrade.gov.vn/en/index.php?option=com_contentandid=759andItemid=76. Accessed 6 March 2016.

Larson, A., Brockhaus, M., Sunderlin, W., Duchelle, A., Babon, A., Dokken, T., Pham, T. T., Resosurdamo, I. A. P., Selaya, G., Awono, A., and Huynh, T.-B. (2013). Land tenure and REDD+: The good, the bad and the ugly. *Global Environmental Change*, 23(3), 678–689.

Lawlor, K., Madeira, E. M., Blockhus, J., and Ganz, D. J. (2013). Community participation and benefits in REDD+: A review of initial outcomes and lessons. *Forests*, 4, 296–318.

Locatelli, B., Evans, V., Wardell, A., Andrade, A., and Vignola, R. (2011). Forests and climate change in Latin America: Linking adaptation and mitigation. *Forests*, 2, 431–450.

McDermott, C. L. (2014). REDDuced: From sustainability to legality to units of carbon – the search for common interests in international forest governance. *Environmental Science and Policy*, 35, 12–19.

McElwee, P. D. (2004). You say illegal, I say legal. *Journal Of Sustainable Forestry*, 19(1–3), 97–135.

McElwee, P. D. (2012). Payments for ecosystem services as neoliberal market-based forest conservation in Vietnam: Panacea or problem? *Geoforum*, 43(3), 412–426.

Ogonowski, M., and Enright, A. (2013). *Cost Implications for Pro-poor REDD+ in Lam Dong Province, Vietnam: Opportunity Costs and Benefit Distribution Systems*. London: IIED.

Patel, T., Dhiaulhaw, A., Gritten, D., Yasmi, Y., Bruyn, T. D., Sharma, N. P., Luintel, L., Khatri, D. B., Silori, C., and Suzuki, R. (2013). Predicting future conflict under REDD+ implementation. *Forests*, 4, 343–363.

Phuc, X., and Nghi, T. H. (2014). *Rubber Expansion and Forest Protection in Vietnam*. Wageningen: Tropenbos International.

Phương, N. (2011). Overview of Lam Dong Province. Available at: www.vietrade.gov. vn/en/index.php?option=com_contentandid=1917:overview-of-lam-dong-province andItemid=275. Accessed 13 March 2016.

Pratihast, A. K., Herold, M., Avitabile, V., Bruin, S., Bartholomeus, H., Souza, C. M., and Ribbe, L. (2013). Mobile devices for community-based REDD+ monitoring: A case study of central Vietnam. *Sensors*, 13, 12–38.

Quang T. N., Luong, T. T, Nguyen, T. H., K'Tip, V., Enters,T., Yasmi, Y., and Vickers, B. (2010). *Evaluation and Verification of the Free, Prior, and Informed Consent Process under the UN-REDD Programme in Lam Dong Province, Vietnam*. Hanoi: RECOFTC (The Center for People and Forests).

RECOFTC (The Center for People and Forests) (2007). People, forests and climate change mitigation – Vietnam: Why REDD+ needs local people. Available at: www. recoftc.org/basic-page/climate-change-mitigation. Accessed 6 March 2016.

Schwarte, C., and Mohammed, E. Y. (2011). *Carbon Righteousness: How to Lever Pro-poor Benefits from REDD+*. London: IIED.

Sikor, T., and Cẩm, H., (2016). REDD+ on the rocks? Conflict over forest and politics of justice in Vietnam. *Human Ecology*, 44, 217–227.

Suhardiman, D., Wichelns, W., Lestrelin, G., and Hoanh, C. T. (2013). Payments for Ecosystem Services in Vietnam: Market-based Incentives or State Control of Resources? *Ecosystem Services* 5: 94–101.

Sunderlin, W. D., Larson, A. M., Duchelle, A. E., Resosudarmo, I. A. P, Huynh, T. B., Awono, A., and Dokken, T. (2013). How are REDD+ proponents addressing tenure problems? Evidence from Brazil, Cameroon, Tanzania, Indonesia, and Vietnam. *World Development*, 55: 37–52.

Thompson, M. C., Baruah, M., and Carr, E. R. (2011). Seeing REDD+ as a project of environmental governance. *Environmental Science and Policy*, 14(2), 100–110.

Thuy, P. T., Moeliono, M., Brockhaus, M., Le, D. N., Wong, G. Y., and Le. T. M. (2014). Local preferences and strategies for effective, efficient and equitable distribution of PES revenues in Vietnam: Lessons for REDD+. *Human Ecology*, 42(6): 885–899.

Thuy, P. T., Moeliono, M., Hien, N. T., Tho, N. H., and Hien, V. T. (2012). The context of REDD+ in Vietnam: Drivers, agents and institutions. *Centre for International Forestry Research, Occasional Paper* 75. Bogor, Indonesia: CIFOR.

To, P. X., Dressler, W. H., Mahanty, S., Pham, T. T., and Zingerli, C. (2012). The prospects for payment of ecosystem services (PES) in Vietnam: a look at three payment schemes. *Human Ecology*, 40(2), 237–249.

To, P., Xuan, L., Mahanty, S., and Dressler, W. H. (2015). 'A new landlord' (địa chủ mới)? Community, land conflict and State Forest Companies (SFCs) in Vietnam. *Forest Policy and Economics*, 58, 21–28.

World Water Assessment Program and UNESCO (2006). *World Water Development Report: Water, A Shared Responsibility*. Paris: UNESCO.

UN-REDD (2009). Reports and analysis: US $4.3 million UN-REDD Viet Nam Programme launched. Available at: www.un-redd.org/NewsCentre/Newsletterhome/US438millionUNREDDVietNamProgrammelaunche/tabid/1469/language/en-US/Default.aspx. Accessed 27 March 2018.

UN-REDD (2013). UN-REDD Viet Nam Phase II Program: Operationalizing REDD+ in Vietnam. Available at: http://vietnam-redd.org/Upload/CMS/Content/REDD%20projects/UN-REDD%20VN%20Phase%202/PD-signed.pdf. Accessed 27 March 2018.

USAID (2011). Climate change in Vietnam: Assessment of issues and options for USAID funding. Available at: www.usaid.gov/sites/default/files/documents/1861/vietnam_climate_change_final2011.pdf. Accessed 6 March 2016.

USAID (2012). Country profile Vietnam – property rights and resource governance. Available at: www.usaidlandtenure.net/sites/default/files/country-profiles/full-reports/USAID_Land_Tenure_Vietnam_Profile.pdf. Accessed 6 March 2016.

Wood, B. T., Stringer, L. C., Dougill, A. J., and Quinn, C. H. (2018). Socially just triple-wins? A framework for evaluating the social justice implications of climate compatible development. *Sustainability*, 10(211), 1–20.

World Bank (2015). The World Bank in Vietnam. Available at: www.worldbank.org/en/country/vietnam. Accessed 27 March 2018.

Zaiha, A. N., Mohd Ismis, M. S., Salmiati, Shahrul Azri, M. S. (2015). Effects of logging activities on ecological water quality indicators in the Berasau River, Johor, Malaysia. *Environmental Monitoring and Assessment*, 187, 493.

7 'We have the right to do anything we like'

Boomerang effects of the Ilısu Dam – lessons to learn

Sonya Deborah Krause, Frances Delaney, Ricarda Ines Konwiarz and Larry Swatuk

Introduction

The Ilısu Dam is part of the South-Eastern Anatolia Project (Turkish: Güneydoğu Anadolu Projesi (GAP)) in Southeast Turkey. GAP is one of the most comprehensive infrastructural projects worldwide, stretching for 75,000 km², about 9.7 per cent of the Turkish territory (Setton and Drillisch, 2006: 2). The GAP currently is composed of 22 dams and 19 hydraulic power plants and irrigation networks. The aims of the project are to improve income levels and quality of life of people living in the region, to reduce regional disparities in Turkey and to contribute to the country's economic and social development (Olcay Ünver, 1997).

The Ilısu Dam is a planned project that intends to dam the water of the Tigris River, which borders the country from Syria and Iraq. The government of Turkey claims that the purpose of the dam is electricity generation. It is estimated that the Ilısu Dam would contribute more than 3800 GWh to the power generation of GAP, making up 16 per cent of the GAP generation and 3.2 per cent of Turkey's overall electricity (Ministry of Foreign Affairs, 2011).

However, this project is associated with many controversies, as the nation frames the project differently than other stakeholders, such as the Syrian and Iraqi governments, the Kurdish population, national and international NGOs and many more. This chapter therefore aims to reveal alternate perspectives on the international, socio-cultural and environmental issues associated with the construction of the Ilısu-Dam in Turkey (Table 7.1), and how these perspectives can reveal or obscure certain issues over others. This chapter also examines possible outcomes and consequences of the state's framing of the dam. These unconsidered consequences can be referred to as local-level side effects and/or boomerang effects (Chapters 1 and 2 in this volume).

Background of the Ilısu Dam

The Ilısu Dam has been shrouded in controversy since its inception. Initial plans for the dam were discussed as early as the 1950s; however, due to

Table 7.1 The international, socio-cultural, and environmental issues associated with the Ilısu Dam, summarized by the Turkish state, and by the other participants in the project

Type	International	Socio-cultural/national	Environmental
State framing	GAP is a peace project	Humanitarian project; Turkish integration	Project enhances environment in barren regions
Opponents' framing	Project precipitates conflict	Human rights offence; project is part of Kurdish suppression	Project destroys environment; brings health hazards

Source: Modified from: Warner, 2011.

disagreements on project design and financing, it was not until 1997 that building was implemented. A consortium, composed of two Swiss companies (Sulzer Hydro and ABB Power Generation), and one British company (Balfour Beatty), was commissioned with the construction of the dam (Setton and Drillisch, 2006: 19). However, because the Ilısu Dam did not fulfil the social and environmental standards of the World Bank and the OECD, and because of international public protests, the first attempt for the realization of the project failed in 2002. In 2004, negotiations were restarted, and a second consortium formed in 2007. Export credit agencies (ECA) from Germany, Austria and Switzerland took on the financial risk of the project. However, in 2008, a report from the governments of the three countries found that Turkey ignored all 153 requirements that had been set by the ECAs in 2007. Again the project was stopped, in 2009. Nevertheless, the Turkish government organized three Turkish banks for the financing of the project. The construction was stopped once more in 2013 because environmental impact assessments (EIAs) were not conducted properly; but, three months later, the Turkish government changed the laws for EIAs and the construction continued (Corporate Watch, 2015).

The Ilısu Dam has also garnered international attention from advocacy groups such as EJOLT (Environmental Justice Organization, Liabilities and Trade) and International Rivers, which have provided support and aid to stop the dam construction from proceeding (Warner, 2012). The fight against the Ilısu Dam has become part of an international movement of civilian voices opposing controversial mega dams, such as the Belo Monte Dam in Brazil. The Ilısu Dam is expected to flood 52 villages and 15 small towns, a total of 12,000 to 78,000 people being affected by the flooding (Eberlein *et al.*, 2010). Missing information regarding the displacement of people in the Resettlement Action Plan (RAP) in Turkey indicates that the majority of the displaced population will be impoverished (Eberlein *et al.*, 2010).

Turkey controls the Tigris and Euphrates river which flows through Turkey into Syria and Iraq, both of which rely on the water from these rivers (Lelieveld *et al.*, 2012). Turkey's control of upstream water and extensive damming practices has created conflict previously. The IPCC predicts that this region of the world will see an increase in droughts, climate variability and fresh water scarcity into the future, affecting downstream agricultural production (IPCC, 2007).

Table 7.2 examines the various stakeholders of the Ilısu Dam. These range from international actors such as Export Credit Agencies and International Rivers, to the communities being displaced at the local level. The scope of this paper does not allow for a detailed analysis of the impact of the Ilısu Dam on all stakeholders. Only the main stakeholders are examined below. Primary stakeholders are those that are directly impacted by the construction of the dam and secondary stakeholders are those that are indirectly affected by the dam.

Environmental dimensions

Environmental impacts

According to the Turkish Government, the construction of the Ilısu Dam is an investment in the best interests of the country and its downstream neighbours, providing energy and food security to all (Warner, 2011). The Tigris River is more flood-prone than the Euphrates and, therefore, the Turkish government has argued that the Turkish dams have been designed to mitigate flooding, and to improve the timing of flow for downstream agricultural needs (Berkun, 2010). In addition, the Turkish Government also states that the dams will aid in times of drought, and will reduce fluctuations in flow of the Tigris River. They believe the dams in the GAP will enhance productivity in downstream farming and guarantee downstream flow (Bilen, 1997). The Turkish government also stresses the benefit of the envisioned increase in irrigable land in Turkey, Syria and Iraq, by 1.7 million ha, 640,000 ha, and 500,000 ha, respectively (Turkey Water Report, 2009). Since agriculture is one of Turkey's most dominant economic sectors, accounting for 20 per cent of the country's GDP, 10 per cent of its

Table 7.2 Overview of stakeholders in the Ilısu Dam (the +/– symbols indicate if the stakeholder will be positively or negatively impacted by dam construction)

	Primary stakeholders	*Secondary stakeholders*
Domestic	• Kurdish (–) • Other local residents; farmers (–) • The biophysical environment (–)	• Residents of Turkey outside the GAP region (+) • Turkish Military (–)
International	• Syrian government (–) • Iraqi government (–) • Turkish State (+/–)	• International advocacy groups (–) i.e. International Rivers, EJOLT • Iraqi civil society groups (–) • Export Credit Agencies (+) • World Bank (+) • Tourists to Hasankeyf (–)

exports and 47 per cent of its civilian employment, the State frames this aspect of the project as extremely valuable (Kaygusuz, 1999).

However, the environmental impacts and cumulative effects that can arise from the construction, river impoundment and dam operation phases of this project are much more destructive than the Turkish government has led people to believe. In this section of the chapter, environmental impacts from the construction of the Ilısu Dam, such as significant reductions in freshwater inflow into Syria and Iraq, increased sedimentation and bedrock erosion, land degradation, increased pollution, increased flooding and a reduction in biodiversity of native species, will be discussed (Warner, 2011; Berkun, 2010).

Increased land irrigation

The extensive amount of agricultural practices that take place in the Southern Anatolia area have already led to increases in salinization and acidification, sedimentation, reduced water and soil quality, and bedrock erosion (Berkun, 2010; Darama et al., 2004). These effects lead to reduced soil quality, and decrease the amount of vegetation cover surrounding the dam, thus increasing the amount of floods in the area (Berkun, 2010). Inappropriate agricultural activities such as over-fertilization, unsuitable irrigation, overgrazing, overexploitation, etc. have caused significant reductions in soil quality, and thus, agricultural yield, in the regions and studies have shown that this is enhanced with increased irrigation use (Olcay Ünver, 1997).

Changes in land use, such as farmlands and urbanization, can significantly change the quantity and quality of water flowing in the Tigris River (Berkun, 2010). For example, increased use of fertilizers will increase the concentration of nitrate and phosphorus in the water, while livestock farms can contribute to higher concentrations of faecal matter in the river (Bach et al., 2002). The increase in impervious surfaces (e.g. roads, roofs, sidewalks, etc.) along a watershed can also alter the water flow regime, as this causes an increase in the rate and quantity of runoff water, and a decrease in groundwater recharge and base flow (Baloch et al., 2015). This therefore will lead to more frequent and larger flooding events, less base flow water draining into streams in periods of drought, more water in lakes and watersheds, and more erosion of river beds and banks, since more bedrock and vegetation will be eroded due to land-use changes and erosion from runoff (Baloch et al., 2015). These impacts will reduce the quality of life for many aquatic organisms, and decrease the natural resources of the entire watershed.

Reduced freshwater quantity and quality

From the construction of the major GAP dams and the significant increase in irrigation use, it is predicted that Syria and Iraq can expect a 47 per cent decrease in freshwater flow from the Tigris River (Kurdish Human Rights Project, 2002). Because of this, Syria and Iraq could experience higher salinity and pollutant content in their freshwater systems from increased usage of fertilization in the area. For example, Kolars and Mitchell (1999) demonstrated that Syria could experience a 35 per cent increase in insecticide levels in the Euphrates River, and another study conducted by the Government of Iraq (2002) showed that Iraq could experience a doubled amount of salinity in the Tigris River from the GAP project. In addition, Iraq believes that the GAP project will impact about 1.3 million ha of agricultural land (40 per cent of all available agricultural land in the country) surrounding the Euphrates and Tigris Rivers, due to the lack of water quality (Government of Iraq, 2002).

Water pollution imposes many environmental and human health impacts, as soil productivity will likely decrease, which can cause previously cultivated land to become barren land. Therefore, due to deteriorated quality of water from increased agricultural use upstream, this can create further water quantity issues in Syria and Iraq, as their available water will not be suitable for agricultural or human consumption (Kurdish Human Rights Project, 2002).

Reduction in biodiversity of native species

Due to the extreme changes in water flow regime of the Tigris and Euphrates Rivers, and from the significant land-use changes associated with the GAP project, there are many biomes and species at risk (Berkun, 2010). The biomes affected include the dry steppe grassland, the oak shrubland, and the woodland of the Mardin mountains. Impacts to these biomes will also have significant impacts on the species that live in these areas, such as the striped hyena and the Caspian tiger (Berkun, 2010).

The Tigris Valley (Hasankeyf and surrounding areas) has also been declared to be a 'Tigris Valley Important Bird Area (IBA)' (Biricik and Karakas, 2012). In this region of Turkey, the Tigris River provides birds with shelter and acts as a natural migration corridor. The area of Hasankeyf also is composed of a unique combination of biomes, as there is a river ecosystem as well as mountain and mountain/steppe habitats, and rocky cliff walls for birds to nest and migrate to in the spring (Biricik and Karakas, 2012). There are 15 bird species that are especially unique to the Tigris Valley, and the construction of the Ilısu Dam could cause irreversible habitat destruction for these birds, which will most likely affect the entire wildlife of the region (Biricik and Karakas, 2012).

Without doubt, there are many significant environmental impacts that may arise, and have already occurred, during the construction of the Ilısu Dam, and the GAP project altogether. The state frames the project as an environmentally beneficial project, as they promise citizens more food and energy security. However, the evidence reviewed here shows clearly that this is not the case. The construction of the Ilısu Dam will cause numerous negative environmental local-level side effects such as increased air pollution, loss of biodiversity, loss in agricultural land, loss in vegetation, and a large loss in water quality. These impacts completely contradict the promises of the Turkish government (e.g. increased food production, flood mitigation, increased vegetation) and have resulted in distinct economic and political boomerang effects negatively impacting the central state's ability to follow through on its intended project (see www.internationalrivers.org/resources/victory-european-governments-backing-out-of-ilisu-dam-project-in-turkey-3535).

Sustainability and climate change

Turkey has been experiencing increased rates of economic, population and industrial growth in the past two decades (World Bank, 2013). Energy demand has increased by 60 per cent from 2002 to 2010, causing Turkey to import expensive energy resources from elsewhere, as the country did not have the means to supply the demand on their own (World Bank, 2013). Imported oil and gas have caused burdens on the country's economy and air quality, as has its commitment to increase thermal power production. This has increased Turkey's total contribution to the world's CO_2 emissions to about 250 million tons (47 per cent from coal, 42 per cent from oil, and 11 per cent from gas) in 2005 (Yüksel, 2010). With increasing energy demand, consumption rates, population and concerns of negative impacts from climate change, Turkey has developed some mitigation and sustainability targets to achieve by 2023 (Blythe *et al.*, 2015), attracting foreign and local private-energy sector investors (Yüksel, 2010). At the turn of the twenty-first century, Turkey claimed to be aiming for reliable, sufficient, timely, economical and environmentally friendly energy sources while providing social development (Kaygusuz, 1999).

By 2023, Turkey aims to have at least 30 per cent of its electricity generated by renewable energy sources (Blythe *et al.*, 2015), and aims to increase this to 100 per cent by 2050, through renewables such as hydropower, wind, geothermal, solar and biomass energies (Turkey's Sustainability Report, 2012). In addition, Turkey aims to reduce its energy intensity levels by 20 per cent between 2011 and 2111 (World Bank, 2013). Turkey has received over $1 billion from the World Bank for renewable energy and efficiency development projects, and thus far these projects have been aiding the country to reduce its CO_2 emissions by three million tons every year (World Bank, 2013). The Turkish state believes that the development of these projects can be seen as a 'triple win' for their country, as domestic energy generation will be increased, CO_2 emissions will be reduced and many of Turkey's energy companies will be expanded. As of November 2017, however, the website ClimateActionTracker (http://climateactiontracker.org/countries/turkey.html) rated Turkey's carbon reduction actions as 'critically insufficient'.

Hydropower from the Euphrates and Tigris Rivers are one of the main renewable resources that Turkey plans on exploiting to meet their renewable resources targets (Yüksel, 2010). About 33–46 per cent of all of Turkey's electric energy demand can be met with hydropower in 2020, and studies show that this can be easily and economically feasible (Yüksel, 2010). Thus, the GAP project was developed to generate 55 billion kWh of electricity per year, which is 45 per cent of Turkey's total hydroelectric potential (Kaygusuz, 1999), with the Ilısu

Dam producing 3800 GWh, or 3.2 per cent of Turkey's energy demand (Ministry of Foreign Affairs, 2011). In 2013, 29 per cent of Turkey's total energy was generated by renewable sources; however, because of their heavy reliance on hydropower, this number decreased to 21 per cent in 2011, as there was less rainfall that year (Blythe *et al.*, 2015).

Climate change thus is an important consideration that needs to be taken into account when developing mitigation strategies and renewable energy projects, as it directly affects precipitation patterns. Precipitation in the Eastern Mediterranean is highly dependent on the synoptic weather conditions and topography in the area (Lelieveld *et al.*, 2012). It has been predicted that during the twenty-first century, climate change could cause a reduction in annual precipitation and increase in mean surface temperatures, affecting atmospheric processes such as evapotranspiration, humidity and soil moisture (Baloch *et al.*, 2015). Climate scientists also predict a northward shift in the jet stream and storm track, causing people living in the Southeastern Anatolia region to rely more on orographic precipitation for water resources (over the Taurus and Zagros Mountains, which flows directly into the Euphrates and Tigris Rivers) (Lelieveld *et al.*, 2012). Other climate models have shown that there may be a significant drying of the entire Eastern Mediterranean–Middle East area, which could cause significant alterations in Turkey's hydrology, and impact downstream food production and water quality (IPCC, 2007). The drying of the land will significantly reduce rainfed agriculture yield, and will reduce the amount of land designated for animal grazing (Evans, 2009). With reduced crop yields, and enhanced drying of the area, vegetation fires are expected to increase, which will decrease air quality in the region (Lelieveld *et al.*, 2012). Climate change impacts are exacerbated when they are combined with increasing pressures from overuse of irrigation from the Tigris and Euphrates River, urbanization, and agricultural activities. By 2040, it is predicted that the demand for fresh water from both the Tigris and Euphrates Rivers for irrigation will far surpass the supply because of these cumulative impacts (Baloch *et al.*, 2015).

Because of the expected reduction in precipitation in the twenty-first century over the Eastern Mediterranean, increased irrigation rates and the changes in atmospheric processes, overall energy production from hydro dams can be reduced significantly (Bekoe and Logah, 2013). Therefore, it is important to consider the uncertainties of the future of the Ilısu Dam, and the other renewable energy opportunities that are available to the country. Hydropower and nuclear power are the two sources of renewable energy that the Turkish government has been focusing on expanding and exploiting, with little attention given to other renewable energy opportunities (Yüksel, 2010). Studies (e.g. Yüksel, 2010; Kaygusuz, 1999) have shown that wind, solar and geothermal renewable energies have the potential to contribute to 8,000 MW, 35 Mtoe (million ton oil equivalent) or 407,050,000 MWh annually, and 35,000 MW of Turkey's total energy, respectively. While there are many legal reforms and incentives for Turkey to shift towards a more sustainable country, there is a lack of coordination between and among all stakeholders, and a lack of transparency in laws and reforms (Blythe *et al.*, 2015). In addition, there is a long bureaucratic process for obtaining licenses for solar and wind projects, inhibiting the development of the desired renewable energy market (Blythe *et al.*, 2015). Therefore, there is a need for centralization of the parties that provide energy-generating licenses in order for Turkey to diversify its energy sources, increase economic development and decrease regional disparities within and outside of the country.

Social controversy

'The Kurdish problem'

From the days of the Ottoman Empire, through the formation of the Turkish state to the present day, the Kurdish population in Turkey has experienced subjugation, cultural suppression, targeted ethnic and gendered violence, displacement, imprisonment and torture. This has resulted in a strong resistance movement from the Kurdish, and the creation of the PKK, the Workers Party of Kurdistan (Partiya Karkerên Kurdistanê in Kurdish). Violent conflict between the PKK and the Turkish state has been ongoing since the late 1970s. The PKK and other rebel groups have been fighting for either separation from Turkey and the formation of an independent Kurdistan, or for more autonomy and political and cultural rights within Turkey without formal separation (see www.bbc.com/news/world-middle-east-33690060 for details). The GAP region in Turkey is inhabited primarily by Kurdish communities.

The GAP project has been rebranded by the state from an irrigation and electrical generation project to a development project intended to bring environmental, social and health benefits to southeast Turkey, following international pressure (Hatem and Dohrmann, 2013). The state discourse of national social and economic development contrasts sharply with the framing of Kurdish groups who argue that the dam construction is a violation of a cease fire, a means to displace 'Kurdistan's people', to break the nation of Kurdistan and to disrupt social solidarity between Kurdish communities living in the region (CSIS, 2016).

Critics have also suggested that the project aims to exert control over the Kurdish population that inhabits the same areas as the GAP project (Jongerden, 2010; Hatem and Dohrmann 2013). For example, the Ilısu dam cuts off routes used by the PKK, and other dams in the GAP region are seen as a means to further militarize the area. The dam will also force Kurdish people to be displaced and thereby weaken social solidarity (Hatem and Dohrmann, 2013). Hatem and Dohrmann (2013) also argue that Turkey frames the GAP project as a means to provide social and economic development, when it is in fact a disguised nation-building project and is using water control as a method of population control. Another dam in the GAP region forces locals to use a centralized, military-controlled ferry, which allows the military to control transportation in the region as well as to closely monitor local activities (Hatem and Dohrmann,

2013). In 2005, the Turkish government stationed 5,000 Turkish soldiers at the construction site of the Ilısu Dam, revealing just how militarized the area had become, serving the dual purpose of both protecting the dam construction and suppressing any Kurdish attempts to assert their sovereignty.

Warner (2011) argues that the Turkish state has replaced a hard power approach with a soft power, hydrological approach to control the Kurdish population and Kurdish independence movements. State authorities have claimed that 'terrorists will no longer be able to easily cross from one region to the other due to the dams', and the Ilısu Dam is especially close to an area with corridors used by Kurdish fighters, which is colloquially named 'Hell's Valley' (Warner, 2011).

This increased militarization, displacement and population control has not come without consequences for Turkey. Perhaps the largest boomerang effect the Ilısu Dam will unleash is the response from Kurdish society against the state. Kurdish rebel groups have claimed responsibility for several recent attacks in Ankara, on military and police personnel. On 17 March 2016, a suicide attack on a military convoy killed 28 people in Ankara. The Kurdistan Freedom Hawks (TAK) claimed responsibility for the attack, citing Turkish security operations in the southeast region of the country, the area in which the GAP is located, as justification (Osborne, 2016). The attack occurred shortly after a curfew was announced for areas in Kurdistan. This attack is just one of many in continued conflict between the Turkish state and the Kurdish population. Dam building further exacerbates this conflict and is used as a means to justify state suppression. The violence that the dam building has caused in the south east is returning to the state in the form of attacks on the military – the very state body that precipitates violence in Kurdish areas of Turkey. The boomerang effect has manifested itself and may continue to manifest itself in the form of decreased security for civil society.

Losing Hasankeyf

The 10.4 billion m³ reservoir of the Ilısu Dam will completely submerge the ancient city of Hasankeyf, one the oldest continuously inhabited cities in the world.[1] The city is the site of an ancient bridge which was used as part of the Silk Road, and is also home to one of the oldest mosques in the world. The loss of this ancient city and the architecture preserved within it has prompted the city to be listed on a watch list of the 100 Most Endangered Sites in the World, by the World Monuments Fund and has caused European funders to withdraw their support.

As mentioned previously, the Ilısu Dam has been rebranded by the state as a project that will bring development to the GAP region, decrease regional disparities and contribute to the overall social and economic development of the country. However, international advocacy groups that oppose the dam point to the historical site as having 'outstanding universal value' with benefits for 'humanity as a whole' (Hodder, 2009). Others, however, have argued that the focus on objects and artefacts decontextualizes heritage (Hodder, 2009; Ronayne, 2007). Ronayne (2007) argues that a focus on artefacts actually promotes the construction of the Ilısu Dam, because advocacy groups are satisfied when monuments are relocated out of reservoir areas. Local campaigners against the Ilısu Dam frame the heritage of Hasankeyf as a site of continuity between the past and present culture (Ronayne, 2007). Campaigners from the Hasankeyf Volunteers said, 'we kept telling them they cannot transport the whole city to some other place. Because Hasankeyf is a whole, you cannot fracture it, you cannot cut it into pieces, there is an organic wholeness there' (Ronayne, 2007). The culture and history that is located in Hasankeyf extends into the recent past for the local inhabitants, though this framing is not well represented (Ronayne, 2007). The flooding of the reservoir would submerge evidence from recent Kurdish conflict, such as village destructions that date back to the 1990s, and possible graves of Kurdish people that were disappeared (Ronayne, 2007). A Kurdish mother who was displaced and whose children were killed in the conflict said, 'Yes, Hasankeyf is our history but the essence of our history was our children. We have no geography anymore, no town and cities. We have nothing left. Our children were our history' (Ronayne, 2007). The framing that Hasankeyf loses its history and value when relocated is one that advocates for more than the preservation of heritage; this framing fully opposes the dam because of the obliteration of the history of Kurdish people, as well as the evidence of state crimes against the Kurdish.

Relocation and resettlement

The building of the Ilısu Dam and the reservoir flooding is expected to displace 78,000 people. This section further examines the discourse of development and opportunity used by the state, which contrasts with the fears of local residents of loss of livelihood (Harris 2008). Further, studies of other relocation projects in the GAP region have found that reservoir flooding and displacement further entrenches gendered inequality and that economic and developmental benefits of reservoirs are not evenly distributed across socioeconomic groups (Harris 2008).

The area affected by the Ilısu Dam is predominantly Kurdish (90%) and the rest are Turkish and Arab (Morvaridi, 2004). The dam would result in the displacement of 184 villages, 85 of which are already abandoned because of previous conflict in the area (Morvaridi, 2004). At least 61,000 people would be displaced, though demographic information from the area is unreliable because of recent displacement and population movement (Morvaridi, 2004). The state framing of the dam as a development project for national benefit is the overwhelming discourse, where the impacts of displacement and the flooding of Hasankeyf are secondary to the benefits. An engineer for the Ilısu Dam said: 'We have to think in terms of not a village or a district or a region but the whole nation. It is the whole country that is going to benefit from this project.' (Morvaridi, 2004). This viewpoint complements the state's framing of the Anatolia

region as being a backward region of Turkey, which can be modernized by building dams and moving ethnic Kurdish people into cities or new settlements
(Morvaridi, 2004). Many local residents, however, frame displacement not as a
means of development, but as a way to assimilate Kurdish people by dispersing
communities and encouraging migration into cities (Morvaridi, 2004). Further,
many farmers work on fertile farms that would be lost under reservoir waters,
and many more do not own land; these people do not benefit from land compensation (Morvaridi, 2004). Experiences from other dam sites have influenced
perceptions of people facing relocation in the planned Ilısu reservoir. For
example, although the state claims to reimburse landowners, up to 37 per cent
of residents living in other flooded areas did not own land previous to flooding
and were not compensated (Warner, 2011). This contributes to local perceptions that resettlement is merely part of a larger state agenda of cultural assimi-
lation (Morvaridi, 2004).

State politics: transboundary impacts of Turkish water politics

The lack of international water law

What makes transboundary water politics often difficult is the fact that there is
little or no universally recognized international water law. In the Tigris basin,
there are several bilateral agreements between the riparian states, such as the
Turkish–Iraqi treaty of friendship and good neighbourliness, and agreements
between Turkey and the two other riparian states, in which Turkey promises a
minimum water flow of $500\,\mathrm{m^3/s}$ to cross the Turkish–Syrian border. However,
these agreements have not ended the tensions among the riparian states.

Furthermore, Turkey was one of the three countries (besides China and
Burundi) that voted against the UN Convention on the Law of the Non-
Navigational Uses of International Watercourses. It has been suggested that
upper riparians tend to believe that the notification process of the Convention
favours downstream riparians by giving them a veto power over projects and
programmes of their states (Salaman, 2007: 9). Indeed, in its water report of
2009, the Turkish General Directorate of State Hydraulic Works states that the
Convention 'goes beyond its scope' because it 'establishes a detailed mechanism
of notification' and 'does not refer to the sovereignty of the states over the parts
of transboundary watercourses located in their territory' (Turkey Water Report,
2009: 50). Hence, Turkey justifies its transboundary water politics with the doctrine of unlimited territorial sovereignty (Warner, 2012: 238). In contrast, Syria
and Iraq usually demand the application of the doctrine of absolute territorial
integrity. As is shown below, because of the lack of a legitimate, overriding
authority, the outcomes of water politics in the basin are mainly determined by
power relationships.

The framework of hydro-hegemony

Zeitoun and Warner (2006) have developed the framework of hydro-hegemony, which accounts for two features: power asymmetries and varying intensities of water conflicts. They suggest that water conflicts can take different forms, ranging from cooperation to violent conflict. Furthermore, they expect that the form of interaction over transboundary waters and, thereby, the scale and dynamics of conflict are chosen by the most powerful state of a basin – the hydro-hegemon (Zeitoun and Warner, 2006: 437). As is shown below, Zeitoun and Warner's framework proves very useful when analyzing the structures and dynamics in the Tigris basin.

Warner (2008; 2011) applies the framework to the Euphrates–Tigris basin. Turkey is considered to be the hegemon of the basin. This status is reasoned historically, since the country was the centre of power of the Ottoman Empire. Furthermore, Turkey has enjoyed long-time membership in and backing by the NATO, and received significant amounts of foreign aid by the United States (Warner, 2008: 281–282). Turkey is also a membership candidate for the European Union. Probably most advantageous is the country's geographical location as the upstream riparian state of the basin.

As is mentioned above, Turkey has posited to cushion floods, to control water flows effectively and fairly, and to adjust flows to the needs to downstream riparians. Except in very dry years, Turkey has kept this promise and created a stability of expectations (Warner, 2008: 280). In other words, Turkey has invested 'great efforts into trying to convince others that the state of affairs is just and legitimate, and that actions are very clearly in the interest of the downstream states as well' (Warner, 2011: 100). However, Turkey's ability to control water flows implies that the county has (political) power over downstream riparian states. Hence, Syria and Iraq do not regard Turkey's dams building exercises as advantageous to the downstream state. Thus, they have protested against any new projects, threatened military action and depicted Turkey's water politics as unilateral and self-serving (Warner, 2012: 246). In fact, what has gotten Syria's and Iraq's most attention is not the real impacts of Turkish dam building, but the *potential* that water flows could either be shut off completely or be released in such amounts to flood downstream territories (Warner, 2004: 17).

However, despite the fears and resistance of the downstream riparians, a violent conflict has never broken out. Warner suggests that, in fact, Syria and Iraq pragmatically accepted the fait accompli and have shown willingness to reconcile and cooperate to a certain extent (Warner, 2012: 246). In 2001, the GAP Regional Development Administration (GAP RDA) and the Syrian General Organisation for Land Development (GOLD) started a cooperation, which aimed for the sustainable utilization of resources through exchange, common projects and training programmes (Scheumann and Kibaroğlu, 2013). In 2005, the Euphrates–Tigris Initiative for Cooperation – a civil society/academic initiative – was established to promote transnational cooperation for development in the basin (Dinar, 2008).

However, despite these limited overtures, it is problematic that such an important shared river basin has no formal, system-wide organization such as the River Basin Commissions that one sees in many other parts of the world (Swatuk, 2017). The outward appearance of cooperation and stability belies the fact of unequal control and distribution of water in the basin. At the same time, Syria and Iraq are heavily dependent on transboundary water resources. According to Shamout and Lahn (2015: 11), approximately 98 per cent of Turkey's water originates within its own borders. In comparison, only 28 per cent of Syria's water and 39 per cent of Iraq's water originate within their own borders. Therefore, according to Zeitoun and Warner's framework, the situation can be categorized as a 'cold conflict'. Such a conflict can become violent if the water stress increases. The framework of hydro-hegemony helps to understand the different framings of the riparian states of transboundary water politics, which is the subject of the next section.

Framing of transboundary water politics

As mentioned above, Turkey emphasizes that each riparian state has a sovereign right over the water resources in its territory, and that Turkish dams and water management have served the interests of lower riparian states as well. A statement of the former Turkish Prime Minister, Suleyman Demirel, illustrates Turkey's perceptions of unlimited territorial sovereignty:

> Neither Syria nor Iraq can lay claim to Turkey's rivers any more than Ankara could claim their oil. This is a matter of sovereignty. We have a right to do anything we like. The water resources are Turkey's; the oil resources are theirs. We don't say we share their oil resources, and they cannot share our water resources.
>
> (KHRP, 2002: 7)

Nevertheless, the country claims that it has consistently abided by 'good neighbourliness' and 'no harm' principles (TWR, 2009, p. 48), and released 'maximum possible amount of water from both rivers even during the driest summers, thanks to the completed dams and the reservoirs in south-eastern Anatolia' (MoFA, 2011a).

As described in the section 'Environmental dimensions', the government makes assurances that the Ilısu Dam is only used for power generation and not for irrigation purposes. Hence, water is rechannelled into the riverbed, whereby water quantities are not reduced. Furthermore, the government argues that the water quality is not reduced but even improved due to new sewage treatment facilities (MoFA, 2011b).

Additionally, Turkey states that it aspires to an 'equitable, reasonable and optimal' use of transboundary water (TWR, 2009: 47; MoFA, 2011a). However, for Turkey such a use does not mean the equal distribution of water among the riparian states. Instead, the state argues that the water in the Euphrates and

Tigris suffices to meet the needs of all riparian states, provided that their use of water is efficient and effective. Turkey demands that *all* riparian states should manage water according to the principle 'more crop per drop,' using appropriate management principles and modern water infrastructures and techniques (TWR, 2009: 48–49). In this regard, Turkey criticizes Iraq and Syria for their inefficient use and waste of water (Al Jabbari *et al.*, 2015). Furthermore, Turkey encourages cooperation as well as knowledge and information sharing among the riparian states. It expects that 'confidence building measures' help to reach solutions in transboundary water issues (TWR, 2009, p. 47).

Although Syria and Iraq had conflicts in the past, the GAP-project has led to their alliance against the 'common enemy.' They have criticized Turkey for 'being arrogant and posting its self-interest as a regional common good' and for not conferring or negotiating with its neighbours (Warner, 2011: 88). When Turkey announced the construction of a new dam, the two countries jointly protested and threatened the country (Warner, 2008: 283). For example, in 1990, Turkey stopped the flow of the Euphrates for nine days to fill the reservoir of the Atatürk dam. Syria and Iraq accused Turkey of not informing them beforehand, and Iraq threatened to bomb Turkish dams (KHRP, 2002). In addition, the two states have claimed that Turkey promised only an *average* downstream flow of $500\,m^3/s$ and that, in some cases, Turkey decided unilaterally and without sufficient notice to release 'extra' water prior to a complete cut-off due to impoundments (KHRP, 2002: 21). Such incidents are a reminder to downstream riparians that Turkey could stop water flows any time. Turkey's three largest dams have a storage capacity that exceeds the entire annual flow of the Tigris and Euphrates, which would allow the complete cut-off of the water flow for considerable time (KHRP, 2002).

Furthermore, downstream riparians have experienced major consequences of Turkey's dam building. Syria has seen two of its smaller rivers run dry. In Iraq desperate farmers have had to abandon their land, and neighbourhoods in Baghdad rely on the Red Cross for drinking water (Al Jabbari *et al.*, 2015). Iraq is particularly concerned about its central marshes, the Middle East's largest wetlands. Half a million people live in these lands and their livelihood depends on the resources of the lands. Around $90\,m^3/s$ would be required for sufficient drinking water availability, irrigation and marsh replenishment, whereas real flows are sometimes as low as $18\,m^3/s$ (Harte, 2013). Furthermore, diminished water levels allow the influx of salt water from the Persian Gulf into the marshes, which is harmful for aquatic life and agriculture (Harte, 2014). Locals see Turkey's dams as being responsible for the increasing hardships they face and for the thousands of 'water refugees' environmental changes cause every year (Harte, 2013).

Despite Turkey's argument that water is channeled back into the riverbed after passing the turbines for power generation, reports of Ilısu opponents have argued that the dam will have significant effects on the hydrology of the Tigris, altering seasonal flow patterns and leading to large fluctuation in daily flows (PWA, 2011; See www.hasankeyfgirisimi.net/?page_id=6 for details). Furthermore, in combination with the Cizre dam, the Ilısu Dam is expected to reduce the flows in Syria and Iraq significantly and 'below historic levels,' maybe even leading to dry riverbeds in the summer (PWA, 2011: 3).

As described above, the last decades have seen limited albeit increasing cooperation among the governments of the riparians. However, civil society organizations in Iraq have started to blame the Iraqi government for a failure to protect its citizens from the potential dangers of the Ilısu Dam and for lacking transparency in transboundary water negotiations (ICSSI, 2013). Iraqi organizations such as Save the Tigris and Iraqi Marshes Campaign (STIMC), the Iraqi Social Forum, and the Iraqi Civil Society Solidarity Initiative have cooperated with foreign organizations, such as the CounterCurrent (GegenStrömung), the Environmental Defender Law Center, the Initiative to Keep Hasankeyf Alive, and The Corner House, in order to raise the voice of Iraqi citizens. In 2015, STIMC sent a representative to the UN in Geneva (ICSSI BAGHDAD, 2015) and submitted a shadow report, which states, among other things, that the Iraqi government failed to take measures to improve access to safe drinking water, to ensure that Turkey and Syria (as well as Iran) release adequate amounts of water into Iraq, and to address the harm that had already been caused by the neighbouring countries (STIMC *et al.*, 2015). In a recent report from the Initiative to Keep Hasankeyf Alive, the authors state that

> In 2017, DSI [State Water Utility] and government officials announced several times that the resettlement of Hasankeyf residents will be done in 2018 and the flooding will start at the end of 2018 or the beginning of 2019.... The Turkish government has declared several times that Iraq would not suffer when the flooding of the dam reservoir takes place. However, until now only 60 m³/s are promised for the flooding period of 6 to 24 months which is very low considering that the average flow is around 500 m³/s at the Turkish–Iraqi border.
>
> (www.hasankeyfgirisimi.net/?page_id=6andprint=pdf)

The high divergence in framing implies that Turkey's water politics have impacts on the downstream states that are, in the best case, unintended, or, in the rather realistic case, simply ignored. These disregarded impacts and the seeming ignorance of downstream voices could lead to the outbreak of conflict if water stresses in Syria and Iraq increase.

The controversy around the Ilısu Dam: 'water war' vs. 'peace project'

The Turkish state posits that it perceives water to be a source for cooperation (although under its own rules and conditions) among the riparian states. As described above, scholars have emphasized that the relationships between the riparian states have improved towards conciliation and peaceful coexistence. The GAP-GOLD agreement and the Euphrates–Tigris Initiative for Cooperation provide examples for steps towards cooperation.

Other authors expect that the construction of the dam could lead to a water war instead (Al-Masri, 2014; Vidal, 2014). In this regard, one major point is that riparians do not really know how much water is there. The water flows show year-on-year differences and the numerous dams lead to further distortions and uncertainties (Jongerden, 2010). A trilateral agreement that ensures close collaboration and information sharing on diversions and flows of water could reduce these uncertainties. However, remaining uncertainties imply vulnerabilities, increasing the potential of violent conflict.

As previously described, factors such as changing agricultural practices, population growth and climate change are expected to increase the water stress in the whole Eastern Mediterranean region. Some have argued that water stress, although it was not the only influencing factor, had a significant impact on creating instability and, thereby, the outbreak of the civil war in Syria (Al Jabbari *et al.*, 2015). Around 1.5 million farmers had been displaced, and because the government did not provide any support, many of them could be recruited by the Islamic State of Iraq and Syria (ISIS) or other groups. The conflict in Syria has also led to the resurgence of the Kurdish issue in Turkey. Armed hostilities and terrorist attacks have led to domestic insecurity in the country. Although there are many other factors involved, one might argue that these developments describe a 'boomerang effect': Turkey posits that its water politics are effective, harmless to downstream riparians, and even cooperative. However, as it ignores consequent downstream developments and complaints by Syria and Iraq, water stresses in these countries contribute to their instability, which in turn affects Turkey's domestic as well as national security. In addition, conflict and destruction in Syria will make transboundary water politics increasingly difficult in the future. Similarly, continuing instability in Iraq have, according to Cascão and Zeitoun (2010: 34), shifted the balance of power in the Tigris–Euphrates basin further in Turkey's favour.

The hydro-hegemony theory helps to understand this controversy around the Ilısu Dam. The theory suggests that water conflict can have different dynamics and outcomes. A violent conflict has not yet broken out and the absence of such a conflict is likely to result from Turkey's strategy to cooperate, integrate and encourage development in order to legitimize its hegemony in the river basin. However, as it has been stated many times by now, the cold conflict could turn into a violent conflict if the water stresses in the basin increase.

Conclusion

This paper brings to light the implications of different framings of water projects, exemplified by the Ilısu Dam. Different framings reveal, obscure and emphasize certain issues over others. The framing of the government obscures social and ethnic stratification, controversies, protests and negative local-level side effects such as the subjugation of the Kurdish population in Southeastern Turkey. The government framing also emphasizes greenhouse gas reduction potential while ignoring other potential detrimental environmental effects of hydropower. The Turkish state, furthermore, claims that its water policies are in the interest of all other riparian states and declares water as 'a catalyst for cooperation' (Ministry of Foreign Affairs, 2011), whereas Syria and Iraq complain about dry riverbeds, farmers who abandon their lands, and Turkey's selfishness. The framework that the government uses exacerbates existing conflicts and blinds the government to potential future issues, which may arise after reservoir flooding, but have been experienced in other reservoir floodings. The issues that state framing obscures do not 'disappear under the water', rather they return to the state in the form of violence from the Kurdish, lessening societal security; in the form of environmental degradation; and in the form of an increased security threat from increasingly water-stressed downstream riparian states. Although the insights shared here will not prevent the Ilısu Dam from being completed, they offer lessons on the issues and contentions that arise in hydro projects. Paying attention to the way that different stakeholders frame hydro projects helps to elucidate potential local-level side effects as well as boomerang effects – undesirable consequences that project planners did not foresee. The social, political and environmental context that dam projects are situated in shape outcomes and the Ilısu Dam and the GAP region show just how destabilizing these outcomes can be.

Note

1 See Eberlein *et al.* (2010) for the extent of the reservoir of the Ilısu Dam, located in the southeast of Turkey, which would flood the city of Hasankeyf and surrounding areas.

References

Al Jabbari, M., Ricklefs, N., and Tollast, R. (2015): Rivers of Babylon. Iraq's water crisis – and what turkey should do. *Foreign Affairs*. 23 August 2015. Available online at www.foreignaffairs.com/articles/iraq/2015-08-23/rivers-babylon.

Al-Masri, A. (2014): Water Wars directed against Syria and Iraq: Turkey's control of the Euphrates River. Turkey's control of the Euphrates might lead to disaster. *Global Research*. 1 July 2014. Available online at www.globalresearch.ca/water-wars-directed-against-syria-and-iraq-turkeys-control-of-the-euphrates-river/5389357, checked on 3 November 2016.

Bach S. J., McAllister, T. A., Veira, D. M., Gannon, V. P. J., and Holley, R. A (2002). Transmission and control of Escherichia coli O157: H7 – a review. *Canadian Journal of Animal Science*, 82: 475–490

Baloch, M. A., Ames, D. P., and Tanik, A. (2015). Hydrologic impacts of climate and land-use change on Namnam Stream in Koycegiz Watershed, Turkey. *International Journal of Environmental Science and Technology*, 12(5), 1481–1494.

Bekoe, E. O., and Logah, F. Y. (2013). The impact of droughts and climate change on electricity generation in Ghana. *Environmental Sciences*, 1(1), 13–24.

Berkun, M. (2010). Hydroelectric potential and environmental effects of multidam hydropower projects in Turkey. *Energy for Sustainable Development*, 14(4), 320–329.

Bilen, Ö. (1997). Turkey and water issues in the Middle East. Southeastern Anatolia Project (GAP) Regional Development Administration.

Biricik, M., and Karakas, R. (2012). Birds of Hasankeyf (south-eastern Anatolia, Turkey) under the threat of a big dam project. *Natural Areas Journal*, 32(1), 96–105.

Blythe, J. W., Uslu, N., and Tosun, I. (2015). Renewable energy level survey: Can Turkey hit the targets. *Turkish Commercial Law Review*, 1, 233.

Cascão, A. E., and Zeitoun, M. (2010). Power, hegemony and critical hydropolitics. In: A. Earle, A Jagerskog and J. Ojendal, eds, *Transboundary Water Management: Principles and Practice*. London: Earthscan, pp. 27–42.

Corporate Watch (2015). Save Hasankeyf, Stop the Ilisu Dam. Available at: https://corporatewatch.org/save-hasankeyf-stop-the-ilisu-dam-2/ accessed 16 May 2018.

CSIS (Center for Strategic and International Studies) (2016). The Kurdish Movement: Politicians and Fighters. Available at: https://csis-prod.s3.amazonaws.com/s3fs-public/publication/160331_kurdish_politicians_fighters.pdf accessed 16 May 2018.

Darama,Y., Hatipoğlu, M. A., Seyrek, K., and Kökpınar, M. A. (2004). Problems related to soil erosion and sediment transport in the Sanliurfa–Harran irrigation scheme. International Congress on river basin management. Ankara.

Dinar, S. (2008). Asymmetry and bargaining power in international environmental negotiations: The case of transboundary water. Paper presented at the annual meeting of the ISA's 49th Annual Convention 'Bridging Multiple Divides' Hilton San Francisco, CA, USA. 26 March 2008.

Eberlein, C., Drillisch, H., Ayboga, E., and Wenidoppler, T. (2010): The Ilısu Dam in Turkey and the role of export credit agencies and NGO networks. *Water Alternatives*, 3(2), 291–312.

Evans, J. P. (2009). 21st century climate change in the Middle East. *Climate Change*, 92, 417–432

Government of Iraq (2002). Position paper indicating Iraq's position on the utilization of the Tigris River Waters. Baghdad.

Harris, L. M. (2008). Water rich, resource poor: Intersections of gender, poverty, and vulnerability in newly irrigated areas of south-eastern Turkey. *World Development* 36(12): 2643–2662.

Harte, J. (2013). In cradle of civilization, shrinking rivers endanger unique Marsh Arab culture. *Explorers Journal*. 24 April 2013. Available online at http://voices.national geographic.com/2013/04/24/in-cradle-of-civilization-shrinking-rivers-endanger-unique-marsh-arab-culture/.

Harte, J. (2014, February 21). New Dam in Turkey Threatens to Flood Ancient City and Archaeological Sites. *National Geographic*. Retrieved from http://news.national geographic.com/news/2014/02/140221-tigris-river-dam-hasankeyf-turkey-iraq-water/

Hatem, R., and Dohrmann, M. (2013). Turkey's fix for the 'Kurdish Problem'. *Middle East Quarterly*, 4, 49–58.

Hodder, I. (2010). Cultural heritage rights: From ownership and descent to justice and well-being. *Anthropological Quarterly*, 83(4), 861–882.

IPCC (2007). *Climate Change 2007 – Impacts, Adaptation and Vulnerability*. Contribution of Working Group II to the Fourth Assessment Report of the IPCC. Cambridge: Cambridge University Press.

Jongerden, J. (2010). Dams and politics in Turkey: utilizing water, developing conflict. *Middle East Policy*, 17(1), 137–143.

Kaygusuz, K. (1999). Energy and water potential of the Southeastern Anatolia project (GAP). *Energy Sources*, 21(10), 913–922.

Kibaroğlu, A., and Scheumann, W. (2013). Evolution of transboundary politics in the Euphrates-Tigris river system: New perspectives and political challenges. *Global Governance: A Review of Multilateralism and International Organizations*, 19(2), 279–305.

Kolars, J. and Mitchell, W. A. (1999). *The Euphrates River and the Southeast Anatolia Development Project*. Southern Illinois University Press, Carbondale.

Kurdish Human Rights Project (KHRP) (2002). Downstream impacts of Turkish Dam construction on Syria and Iraq: Joint report of fact-finding mission to Syria and Iraq. The Corner House. Available online at www.thecornerhouse.org.uk/sites/thecorner house.org.uk/files/IraqSyri.pdf.

Lelieveld, J., Hadjinicolaou, P., Kostopoulou, E., Chenoweth, J., El Maayar, M., Giannakopoulos, C., and Xoplaki, E. (2012). Climate change and impacts in the Eastern Mediterranean and the Middle East. *Climatic Change*, 114(3–4), 667–687.

Ministry of Foreign Affairs (2011). Turkey's policy on water issues. Republic of Turkey Ministry of Foreign Affairs. Available online at www.mfa.gov.tr/turkey_s-policy-on-water-issues.en.mfa, updated on 2011, checked on 3/11/2016.

Morvaridi, B. (2004). Resettlement, rights to development and the Ilısu Dam, Turkey. *Development and Change*, 35(4), 719–741.

Olcay Ünver, I. H. (1997). Southeastern Anatolia Project (GAP). *World Resources Development* 13(4): 453–483.

Osborne, S. (2016). Ankara bombing: Kurdish militant group claims attack in Turkish capital. *Independent*. Retrieved from: www.independent.co.uk/news/world/middle-east/ankara-bombing-kurdish-militant-group-claims-attack-in-turkish-capital-a6936031.html#gallery

Philip Williams and Associates (PWA) (2001). A review of the hydrological and geomorphic impacts of the proposed Ilısu Dam. Report for The Corner House. California. Available online at www.thecornerhouse.org.uk/sites/thecornerhouse.org.uk/files/ilisueia.pdf.

Ronayne, M. (2007). The culture of caring and its destruction in the Middle East: Women's work, water, war and archaeology. In: Y. Hamilakis and P. Duke, Eds, *Archaeology and Capitalism: From Ethics to Politics*. Walnut Creek, California: Left Coast Press, pp. 247–265.

Salman M. A. (2007). The United Nations Watercourses Convention ten years later: Why has its entry into force proven difficult? *Water International*, 32(1), 1–15.

Save the Tigris and Iraqi Marshes Campaign (STIMC), Iraqi Social Forum, Iraqi Civil Society Solidarity Initiative, CounterCurrent – GegenStrömung/INFOE, Environmental Defender Law Center, Initiative to Keep Hasankeyf Alive, The Corner House (2015): The Ilısu dam and its impact on human rights in Iraq: The Iraq government's failure to act. Submission to the UN Committee on Economic, Social and Cultural Rights for its 56th Session, 21 September–9 October 2015. Available online at www.iraqicivilsociety.org/wp-content/uploads/2015/09/CESCRIraq_Submission-by-Save-Tigris-Campaign-1.pdf, checked on 4/13/2016.

Setton, D., and Drillisch, H. (2006). *Zum Scheitern verurteilt. Der Ilısu-Staudamm im Südosten der Türkei*. Berlin: World Economy, Ecology and Development (WEED).

Shamout, M. N., Lahn, G. (2015). *The Euphrates in crisis: Channels of cooperation for a threatened river*. Research Paper. London: Chatham House.

Swatuk, L. A. (2017). *Water in Southern Africa*. Pietermaritzburg: UKZN Press.

The Iraqi Civil Society Solidarity Initiative (ICSSI) (2013). Press Release: Include Ilısu dam on the meeting agenda with Turkish Opposition Leader. 20 August 2013. Available online at www.iraqicivilsociety.org/archives/2278, checked on 4/13/2016.

The Iraqi Civil Society Solidarity Initiative Baghdad (ICSSI BAGHDAD 2015): 'Save the Tigris and Iraqi Marshes': The Threat of Ilısu Dam. The Failure of the Iraqi Government to Protect Iraqi's Right to Water. *Global Research*. Available online at www. globalresearch.ca/save-the-tigris-and-iraqi-marshes-the-threat-of-ilisu-dam/5479148, checked on 6 October 2018.

Turkey Water Report 2009 (2009). General Directorate of State Hydraulic Works. Ankara.

Turkey's Sustainable Development Report: Claiming the Future (2012). Ministry of Development. Ankara.

Turkish Ministry of Foreign Affairs (MoFA) (2011a). Turkey's policy on water issues. Republic of Turkey Ministry of Foreign Affairs. Available online at www.mfa.gov.tr/ turkey_s-policy-on-water-issues.en.mfa, updated on 2011, checked on 13 April 2016.

Turkish Ministry of Foreign Affairs (MoFA) (2011b). Ilısu Dam. Available online at www.mfa.gov.tr/ilisu-dam.en.mfa, updated on 2011, checked on 13 April 2016.

Vidal, John (2014). Water supply key to outcome of conflicts in Iraq and Syria, experts warn. Edited by *Guardian*. 2 July 2014. Available online at www.theguardian.com/ environment/2014/jul/02/water-key-conflict-iraq-syria-isis, checked on 11 March 2016.

Warner, J. (2004). Plugging the GAP. Working with Buzan: the Ilısu Dam as a security issue. Occasional Paper No 67.

Warner, J. (2008). Contested hydrohegemony: Hydraulic control and security in Turkey. *Water Alternatives*, 1(2), 271–288.

Warner, J. (2011). *Flood Planning: The Politics of Water Security*. New York. IB Taurus and Co.

Warner, J. (2012). The struggle over Turkey's Ilısu Dam: domestic and international security linkages. *International Environment Agreements*, 12(3), 231–250.

World Bank. (2013). Wind, water, and steam – a triple win for Turkey's energy sector. Retrieved on 12 April 2016, from: www.worldbank.org/en/news/feature/2013/05/30/ wind-water-steam-a-triple-win-for-turkey-energy-sector

Yüksel, I. (2010). Energy production and sustainable energy policies in Turkey. *Renewable Energy*, 35(7), 1469–1476.

Zeitoun, M., and Warner, J. (2006): Hydro-hegemony – a framework for analysis of trans-boundary water conflicts. *Water Policy*, 8 (5), 435. DOI: 10.2166/wp.2006.054.

8 Can climate change challenges unite a divided Jordan River Basin?

Nikita Yasmin Shah, Lars Wirkus and Larry Swatuk

Introduction

Transboundary water management and cooperation remains a difficult issue within the Middle East, particularly among the riparian states of the Jordan Basin. Environmental changes, inclusive of climate change, profoundly impact the region, thereby affecting water and food security and the desire of riparian states to be self-sufficient. In its Global Risks Report 2018 (WEF), the World Economic Forum identified water crises and the failure of climate change mitigation and adapted them as two of the top five global risks in terms of impact. The water crisis still remains the most predominant risk in the Middle East as already laid out in their Global Risks Report 2016 (WEF, 2016). Riparian states remain vulnerable to several existing and foreseeable threats including but not limited to forced migration, displacement, human rights violations, high unemployment and stagnated development. The World Bank indicates that the region's growth rates may decline by as much as 6 per cent of GDP by 2050 due to water scarcity (World Bank 2016c). When resources are scarce, it is well known that countries are prone to conflict; the riparian states are no different. Whilst there have been persistent small-scale conflicts, glimmers of cooperation can be found in negotiations and agreements made between riparian states (Allan, J. A., 2002), which over time have revealed Israel as the hegemon of the basin (Zeitoun, 2008) rendering Jordan, Syria and Lebanon limited access to the basin and Palestine no access at all. As water remains a contentious topic, amongst other issues within the ongoing conflicts, the basin itself is highly politicized and securitized; therefore, addressing issues of water allocation and management directly is high politics. In this chapter we hypothesize that by actions towards tackling climate change and reducing vulnerability to environmental change through mitigation and adaptation will help to address the Israel–Palestine conflict, reduce regional tensions and build habits of cooperation.

In terms of climate change, the region's states have responded individually through the Paris Agreement and COP21. Given water's central role in society's ability to adapt to climate change, it is clear that Jordan River Basin states must cooperate on sustainable water resource management if long-term security is to be achieved. So, the evidence presented here shows that far from cooperating,

Israel prefers to present itself as a technological innovator in the service of adaptation and mitigation. At the same time, the entire region remains highly dependent on the importation of staple foods, that is, what Allan (Allan, J. A., 2002) defines as virtual water.[1]

In our view, virtual water is at best a short-term solution. At worst, it creates the policy space necessary for politicians to pursue business as usual, so leading towards climate-related regional insecurity. This chapter builds on Allan's (Allan, J. A., 2003; 2002) argument that virtual water enables policymakers and politicians to behave in a manner that ignores the prominent issue of the water–food–energy–climate nexus. Virtual water gives the illusion of sufficient water and food and downplays the reality of water scarcity. The effects of climate change are worldwide, and Israel is aware of its role as a leader in desalination technology, placing Israel in a strategic and performative position. On the one hand, Israel can securitize water resources; take mitigation and adaptation for climate change actions, whilst preserving their hegemonic position in acquiring more power to establish national water security, encouraging other global actors to perceive Israel as a respectable and noble actor on the world stage. However, establishing such a position enables Israel to keep enclosed its ongoing conflict with Palestine, where Israel partakes in violating human rights and illegal settler activity, so ensuring a divided and conflictful region.

There are three foreseeable negative local-level side effects:

- As Israel continues the development and installation of desalination plants, the technology will negatively impact biodiversity and the environment. Furthermore, cost of goods produced using desalinated water are higher because the cost of desalinated water is higher than water from other sources thereby making the goods unaffordable or inaccessible for those with low incomes, so exacerbating social inequalities.
- Dependence on virtual water increases the risk of commodity volatility so leading to social instability as well as land and water grabs within Palestine.
- The human rights of Palestinians will continue to be violated.

In turn, there will be no challenge to the status quo, maintaining Israel's position as the hegemon who can thus enact policies and procedures to securitize its own water and food security in the future at the expense of other riparian states.

The Jordan Basin

The Jordan Basin has five riparian states: Israel, Jordan, Lebanon, Syria and Palestine (Gaza Strip and the West Bank). The Basin is made up of many smaller freshwater lakes and rivers and is predominately a semi-arid area.

The World Resource Institute in 2015 evaluated global water stress in 2040 under a business as usual scenario assessing agriculture, domestic and industrial sectors. Results indicated that riparian states of the Jordan Basin scored between four and five, placing them all in the 'extremely high' level of water stress category (Luo *et al.* 2015, p. 6). The projected change in temperature increases the likelihood of droughts, desertification and predominantly changing conditions for agriculture that are further stressed by the decrease in rainfall projections (World Bank, 2016a, 2016b).

(Water) conflict history of the Jordan Basin

The Jordan Basin has a long and complex history involving violence and conflict. Whilst water has not always been at the heart of these conflicts, it has been used as a strategic resource for states to obtain greater power and ensure security. The basin has become highly politicized and securitized as it continues to deal with the colonial legacies of the Ottoman and British Empire, current occupation of Palestine by Israel, unequal distribution of resources between upstream and downstream states and asymmetric power relations, making it difficult to achieve peace through negotiations and cooperation. The nature of a transboundary source is such that when a state attempts to divert water and or build infrastructure, the supply of water for other riparian states is affected (Weinthal *et al.*, 2015: 2). Nonetheless, riparian states are motivated to find solutions to their water issues independently or by partaking in bi- or multilateral negotiations. As water remains deeply political and integral to nation-building, a shared solution for the transboundary resource in this ideologically divided region seems unlikely to be realized any time soon.

Conflict over water in the Jordan Basin can be traced back to the creation of the State of Israel, which began not in the Middle East but in Europe. Anti-Semitism, oppression and discrimination against Jewish people led to the Zionist movement – to develop and foster a nation for Jews in historic Palestine, led by the Zionist myth, a land without people for a people without land (Habib 2007: 1099). The British military occupied Palestine in 1917–1918 and later ruled in 1918–1948. Britain declared their support for a Jewish state in the Land of Palestine in the Balfour Declaration in 1917 (IMFA 2016). Later, in 1922, the League of Nations issued the British Mandate for Palestine, marking formal British rule and the period in which Britain assisted the Zionist movement until the end of the Mandate in 1948. Israel declared independence after the end of the British Mandate, marking this as a moment of celebration for Jews and 'Al Nakba' (the catastrophe) (Al Jazeera 2013) for Palestinians; over 750,000 Palestinians were expelled from their homes and forced into exile.

In 1951, Jordan made public plans to irrigate the Jordan Valley using the Yarmouk River for agriculture and development. Israel responded by initiating the drainage of the Huleh swamps situated between Israel and Syria's demilitarized zone, causing a dispute over borders (Wolf, 1997 and Samson and Charrier, 1997 in Gleick, 2004: 240). In 1953, Israel began constructing the 'National Water Carrier' for irrigation purposes; water would be transferred from the North of the Sea of Galilee out of the Jordan Basin and into the Negev Desert (Samson and Charrier, 1997 in Gleick, 2004: 242). However, due to international disapproval and Syrian military actions along the border, Israel had to move its intake to the Sea of Galilee (Samson and Charrier, 1997 in Gleick, 2004: 242). The Arab League gathered at the Arab Summit in 1964 to discuss plans on halting Israel's National Water Carrier and to divert the headwaters Hasbani and Banias of the Jordan River to Syria and Jordan (Wolf, 1995, 1997 in Gleick, 2004: 243; FAO Aquastat 2009: 9). Upon Israeli's learning of discussions over the new diversion projects from which Israel was excluded, Israel attacked diversion projects in Syria from 1965 to 1967. These small-scale conflicts, alongside others, led to the Six-Day War in 1967 that resulted in Israel gaining control over the entire Lower Jordan River, Mountain Aquifers in the West Bank, Gaza Strip and Golan Heights, dramatically changing the power relations within the basin. Israel had become the most powerful riparian state and henceforth used this power to dominate over riparian states and to pursue national building. Israel began to confiscate and/or destroy Palestinian irrigation pumps on the Jordan River, introduced quotas on existing Palestinian irrigation wells, forbidding the establishment of new ones, and prevented Palestinians from accessing and using water from the Jordan River; instead, water was to be purchased from Israel, establishing a relationship of dependency and control whilst disregarding Palestinian water rights (FAO Aquastat 2009: 9). Israel established settlements in Palestinian areas (established in the Armistice Agreement), where water continues to be unequally distributed amongst Palestinians and Israeli settlers. In 2001, Palestinians destroyed water-supply pipelines connected to Israeli settlements in the West Bank; thus Israel responded by disconnecting the Aqbat Jabar refugee camp from its water supply pumps (ENS, 2001 in Gleick, 2004: 249). To this day, Israel cuts water supplies as a punishment or to exert control. During Ramadan in 2016 Israel cut water supplies in the municipality of Jenin in Palestine, preventing Palestinians, including those who were fasting, from collecting water (Yeung 2016; Al Jazeera 2016). Such actions hinder peace and incite conflict.

In 1978, Israel invaded Lebanon and seized control of the Wazzani spring/stream feeding Jordan, limiting Lebanon's access (FAO Aquastat, 2009: 9). Disputes between Israel, Syria and Lebanon led to Israel cutting off Beirut's water supply in 1978, causing the economy to disintegrate and hindering development processes (Wolf, 1997 in Gleick, 2004: 245). In 2002, tensions deepened between Israel and Lebanon when Lebanon announced the construction of a new pumping station at the Wazzani springs, which would reduce the water levels in the Sea of Galilee affecting Israel's water supply (Luft, 2002). Israel's acquisition of the water source has largely impacted development in the area and increased the vulnerability of inhabitants; tensions remain tense.

(Water) cooperation history of the Jordan Basin

Through these accounts of conflict, tensions over water and struggles over power are evident. At the same time, some agreements have been negotiated. In 1953, the United States sent Ambassador Eric Johnston to aid negotiations on the allocation of the Jordan River for irrigation use, which later became known as the Johnston Plan. Through several rounds of negotiations, $720 \times 106\,m^3$ per year was allocated to Jordan, $400 \times 66\,m^3$ per year to Israel and $132 \times 106\,m^3$ per year to Syria (the figure for Lebanon varies and the concrete allocations have been disputed) (Phillips et al., 2007 in Bismuth, 2016: 197). Participating states did not formally ratify the plan, as they believed the US did not remain impartial during the process, yet these allocations are often referred to as a baseline for future negotiations or projects. The involvement of the United States in these negotiations reflects the pursuing of their own interest in establishing a presence in the region and the special relationship between Israel and the US (Kally and Fishelson, 1993: 32–33). Moving forward in time, the United States continues to financially and politically support development in Israel, making concrete the status of Israel on the global stage. Though the Johnston plan is often cited as the basis for cooperation, cooperation has neither been equal nor in the best interest of all riparian states, rather cooperation remains in the best interest of Israel and Zionism (Kally, 1993: 32–33). Furthermore, cooperation was conceptualized using a Realist mindset allowing for the securitization of resources for Israelis and the state of Israel (Allan, J. A., 2002: 255–256).

Another significant agreement is the Peace Treaty signed by Jordan and Israel in 1994, where water issues regarding the allocation of water from the Yarmuk and Jordan rivers were settled using the principles of international customary water law; the rule of equitable and reasonable utilisation, the no-harm rule and the duty to cooperate. Article 6 in Annex II allocated Israel $12 \times 10^6\,m^3$ from the Yarmuk River, and the remainder to Jordan in the summer period. In the winter period, Israel was allocated $13 \times 10^6\,m^3$; Israel can pump an extra $20 \times 10^6\,m^3$ as Israel transfers $20 \times 10^6\,m^3$ to Jordan from the Jordan river in summer (Treaty of Peace, 1994). The agreement states that 'Jordan is entitled to an annual quantity equivalent to that of Israel, if Jordan's use will not harm the quantity or quality of the above Israeli uses' (Treaty of Peace, 1994). Article 7 of Annex II establishes that three members of Jordan and Israel each will constitute the Joint Water Committee (JWC) whose responsibilities include monitoring water use, preventing harm and managing water issues between Israel, Jordan and Palestine (Treaty of Peace, 1994). The agreement is skewed in the favour of Israel and fails to address a crisis whereby either Israel or Jordan need more water; if Jordan goes beyond its allocation then Israel's allocation could be harmed, leading to Jordan receiving a sanction. Whereas, in the reverse situation Israel can do so in accordance with the agreement, due to their geographic positioning in the basin; it is difficult to accurately measure and keep to these allocations. Though the agreement demonstrates cooperation, it does not factor in sustainability nor the impacts of climate change.

The Oslo II agreement deals with the use of aquifers in the West Bank, the Eastern, Western and North-Eastern Mountain Aquifers, between Israel and Palestine. Annex III, Article 40, settled the distribution of water and established the Palestinian Water Administration Authority, which recognized Palestinian water rights for the first time. Despite such progress, the agreement acknowledges the involvement of Palestinians at Israel's discretion. The JWC perpetuates inequalities amongst riparian states; though decisions are to be mutually agreed upon, Israel alone has a de facto veto right. Bismuth (2016: 198–199) highlights that 'projects outside the areas under the administration of the Palestinian Authority (A and B) need the approval of the civil administration which represents a branch of the Israeli Defence Ministry', hence removing Palestinian agency to make decisions and preventing Palestinians in moving forward in developing their water resources until they are granted permission, often prolonging the process. Developments in Palestine should reflect the needs of Palestinians, but by shunning Palestinian involvement and contribution, local knowledge is not involved in the process. Palestine also moves strategically and blocks any plans proposed by Israel to improve settlements in the West Bank (Bismuth, 2016: 199). It is likely that failure to cooperate from Palestine's side is perceived as negative and as though Palestine is unwilling to cooperate with Israel towards achieving peace. This narrative ignores the simple fact that Israel will only cooperate when Israel benefits most. By maintaining the status quo, Israel too is hindering peace processes.

Population increase and migration as neglected drivers of water insecurity

Though negotiations and agreements give an indication of water allocations for each state, they fail to account for the added stress on resources caused by an increase in population due to mass migration or conflict. With the support of Israelis within Israel and the diaspora, the population of Israel continues to grow, resulting in an increase of illegal settlements in areas such as the West Bank where land and water resources are already not enough for Palestinians. Despite constraints on water, Israel pursues a 'sanctioned discourse' narrative; needing more water to secure and sustain the livelihoods of Israelis and to justify settler activity in an already densely populated area, pitting the needs of Israelis against Palestinians. The West Bank has a population of around 2.6 million people, of which more than 700,000 are Palestinian refugees; 500,000 Israeli settlers also live in the West Bank territory and the disputed areas of East Jerusalem (Meisen and Tatum, 2011: 10). The Gaza Strip has a population of around 1.7 million people; over one million people are Palestinian refugees (Meisen and Tatum, 2011: 10). With conditions worsening for Palestinians, forced migration continues onward into neighbouring states, with currently more than two million Palestine refugees in Jordan, 526,000 in Syria and 450,000 in Lebanon (IRIN, 2018), increasing the population, adding stress on resources and increasing the precarity of those that move. Over the past years, wars and violent conflicts in Syria have also caused mass migration into Lebanon, increasing the population by 30 per cent with the entrance of more than 990,000 Syrian refugees and more than 600,000 Syrian refugees into Jordan (UNHCR, 2018; UNFCCC, 2015c: 2).

Water allocations are usually calculated to ensure that citizens have access to enough water to meet their needs and for the states to develop. Although also refugees and internally displaced peoples require access to good quality water in sufficient amounts for food, sanitation, development and dignified livelihood, they are, however, not always a part of the calculations in a meaningful manner. The surge of population in Lebanon due to the arrival of the Syrian refugees between 2012 and 2014 alone has led to an increase in domestic water demand for refugees, equivalent to a 12 per cent increase of the national water demand (UNFCCC, 2015c: 2). The absence of dialogue around refugees and their needs suggests that their right to water is not focused on, and when attention is raised regarding refugees and resources, the narrative rather implies that refugees and internally displaced people (IDPs) are perceived as a burden, and the cause of energy-production challenges and increased greenhouse gas emissions. As of 2015, riparian states Israel, Jordan and Lebanon signed the Paris Agreement to combat climate change consequences.

COP21 and the NDCs

The Paris Agreement came into force on 4 November 2016, where Parties committed to taking actions to combat climate change and adapt to its effects (UNFCCC, 2016). The Paris Agreement's central aim is to 'strengthen the global response to the threat of climate change by keeping a global temperature rise this century well below 2°C degrees Celsius above pre-industrial levels and to pursue efforts to limit the temperature increase even further to 1.5°C' (UNFCCC, 2016).

Development has led to the increase of greenhouse gas (GHG) emissions, and although riparian states are not the largest contributors to these emissions, Israel, Jordan and Lebanon have assessed what their projected contributions will be and have made the following commitments:

- Lebanon contributes 0.07 per cent global emissions and has committed to a

    GHG emission reduction of 15 per cent compared to the BAU scenario in 2030, additionally to generate 15 per cent of power and heat demands in 2030 using renewable energy and to implement energy efficient to reduce power demand by 3 per cent in 2030 under a BAU scenario.
                                                                    (UNFCCC, 2015c: 6)

- Jordan contributes 0.06 per cent of total greenhouse gas emissions, which roughly equates to 28,717 Gg of CO_2 eq; these emissions are likely to grow to 51,028 Gg of CO_2 eq in 2030 (UNFCCC, 2015b: 2). Jordan has committed to reduce GHG emissions by 14 per cent; a 1.5 per cent reduction under a BAU scenario by 2030 (UNFCCC, 2015b: 1).
- Israel contributes 0.2 per cent of global greenhouse gas emissions and has committed to 'reduc[ing] its per capita greenhouse gas emissions to 7.7 tCO_2e by 2030 which constitutes a reduction of 26 per cent below the level in 2005 of 10.4 tCO_2e per capita' (UNFCCC, 2015a: 1).

To meet these commitments and to outline adaptation to climate change impact plans, each state submitted a 'Nationally Determined Contribution (formally known as Intended Nationally Determined Contributions prior to the enforcement of the agreement) wherein goals, plans and targets were outlined. Jordan's NDC was the most detailed, highlighting their knowledge, commitment and extensive existing and future programmes. Lebanon's NDC was less extensive, yet still outlines adaptation and mitigation plans, though it is clear that if riparian states work together, Lebanon could benefit from knowledge transfer.

Israel's NDC

Israel considers technology, most specifically desalination and drip irrigation, as its best contribution to climate change adaptation and mitigation, which will leverage Israel as a centre of knowledge at a regional and global scale (IMEP, 2009: 7–8). Due to previous experiences of severe drought, Israel has developed innovative technologies for water management, manufacturing, recycling and reuse of treated wastewater, desert agriculture and afforestation. As Jordan and Lebanon seek to also improve in these areas, Israel has a powerful role in aiding the transfer of knowledge and technology; hence it is in the interest of Jordan and Lebanon to cooperate with Israel to reap the benefits. Dr Yeshayahu Bar-Or, Chief Scientist of the Ministry of Environmental Protection, strongly believes that 'in about 20 years, Europe will experience some of the conditions that Israel faces today' (IMEP, 2009: 7–8). Minister Avi Gabbay believes that Israel's technology can create 'business opportunities for investors and for markets, related to the fields of energy efficiency, renewable energies, and alternatives to polluting fuels' (IMEP, 2016: 5). Furthermore, Gabbay believes that Israel can play a significant role in helping the world reach the targets that were agreed upon in Paris (IMEP, 2016: 5). Hence, Israel is positioning itself in a place of future power.

Jordan's NDC

Jordan has provided an extensive list of adaptation and mitigation actions. Here we focus on a few within the agriculture and water sectors. Jordan has identified rural areas as most susceptible to the impacts of climate change as they are most dependent on agriculture and spent more of their income on agriculture (UNFCCC, 2015b: 15). Hence, Jordan seeks to implement sustainable agriculture policies and to develop agronomic and crop strategies to prevent a loss of productivity from climate change impacts and to build resilience amongst farmers to better deal with socio-economic issues such as crop diversification, to include both food and cash crops, to increase the income of farmers, modifying cropping patterns and calendars improving planting and harvesting times and aligning these times as the climate alters (UNFCCC, 2015b: 15–16). Jordan aims to produce crops that are less water-intensive, more tolerant to droughts and salinity (UNFCCCb, 2015b: 15–16). To improve water efficiency, Jordan will introduce drip irrigation, micro-spray and night irrigation to reduce water loss through evaporation and minimize water waste (UNFCCC, 2015b: 16) (also a part of mitigation actions (UNFCCC, 2015b: 7–8). To increase water availability, Jordan, like Israel seeks to manufacture water using desalination technology for agricultural use (UNFCCC, 2015b: 16). Lastly, to emphasize water efficiency and reduce water waste, Jordan will reform water pricing to assert the value of water (UNFCCC, 2015b: 15–16). Jordan's adaptation planning extends over into socio-economic issues and acknowledges the importance of raising awareness about water-related issues and preparing community members of all ages and social statuses, teaching everyone how to better save water and improve sanitation methods and introducing policy measures to ensure the equity in access to water (UNFCCC, 2015b: 12), though these policies are directed at citizens and fail to explicitly mention refugees.

Lebanon's NDC

Lebanon aims to minimize water loss to adapt to climate change impacts, though has not stated how this goal will be achieved (UNFCCC, 2015c: 4). To increase water availability, groundwater aquifers, surface storage dams and hill lakes will be increased through artificial recharge (UNFCCC, 2015c: 4). Lebanon seeks to increase wastewater collection and treatment to improve sanitation and for its reuse after treatment (UNFCCC, 2015c: 4). With regard to agriculture, Lebanon wants to improve 'water efficiency and decrease water loss in irrigation (UNFCCC, 2015c: 4). Also, Lebanon aims to rehabilitate existing water resource networks and install water metres to better understand water use (UNFCCC, 2015c: 4).

Technological solutions to water security

Desalination

Israel manufactures water using 'seawater reverse osmosis' at desalination plants located in coastal towns of Sorek (Soreq), Ashkelon, Hadera, Ashdod and Palmachim. Permeate water can be, and is, used for sanitation, drinking, industry, domestic use and agriculture, including drip irrigation, as the lowered levels of saline will not damage crops, thereby relieving other water sources such as the Dead Sea from over extraction or exploitation (Dreizin et al., 2007: 6).

Water levels in the Dead Sea have declined steadily at a rate of one metre per year due to the over-extraction of groundwater for agricultural purposes (Bismuth, 2016: 94). To combat the decline, Jordan and Israel established the Red Sea-Dead Sea Conveyance (RSDSC) project in association with the World Bank and the Palestinian Authority. In December 2013, Minister Silvan Shalom for Israel, Minister Hazim El-Naser for Jordan and Minister Shaddad Attili for the Palestinian Authority signed a Memorandum of Understanding (MoU) (World Bank, 2013a). The new desalination plant is to primarily manufacture water to be shared between Israel and Jordan and for Merekot, an Israeli water utility, to sell around 20 million cubic metres per year to the Palestinian Authority to be used in the West Bank for agricultural and domestic purposes (World Bank, 2013b). Second, the project seeks to divert the brine produced from desalination to the Dead Sea to increase water levels.

Although desalination can have many advantages – producing water, encouraging cooperation, restoring a symbolic water source – it can also have significant repercussions on the environment, including, but not limited to, affecting marine environment, chemically altering the balance of seawater and increasing energy consumption to run the plants (Einav *et al.*, 2002: 142–143). Additionally, there are social and environmental implications, including, but not limited to, the creation of noise pollution, and land use, installing infrastructure and its economic impact on tourism and goods produced using desalinated water (Einav and Lokiec, 2003: 80).

Also, the location of desalination plants matter. For one, they must be close to the sea and the shore, yet far enough to prevent disruption in society. A desalination plant that produces 100 million m³/year requires 25 acres of land (Einav *et al.*, 2002: 145). Plants are often large, noisy and require a vast amount of energy. Where land is sparse, each metre is of great value; areas that could have been used for internal or external tourism are instead used to build these plants. Additionally, future desalination plants may be built on areas of contention – it is important to evaluate the impacts on Palestinians, particularly located in the Gaza Strip with regard to land use.

Drip irrigation

Both Jordan and Lebanon seek to adopt drip irrigation for agricultural production and to improve water efficiency and reduce waste. Drip irrigation 'minimizes the use of water and fertilizer allowing water to drip slowly and precisely to the roots of plants, either onto the soil surface or subsurface – directly onto the root zone' (Netafim, 2015). Despite efficiency, drip irrigation remains underused by farmers due to lack of knowledge, implementation costs, lack of access to technology and tools and other impediments. Gaza primarily uses intensive irrigated farming whereas the West Bank primarily utilizes rain-fed farming. Syria, Lebanon and Jordan use a variety of methods, including rain-fed, surface irrigation, flood irrigation, re-use of treated wastewater, micro-spray and more to grow an array of crops.

Crop production and export commodities of riparian states

Israel, Jordan and Lebanon all export high-value crops such as chillies, citrus fruit, olives, dates and tomatoes. Root vegetables and a variety of other fruits such as bananas are also exported. In addition, Lebanon also produces sugar and wheat for export (FAOSTAT, 2016). Many of these crops are highly water intensive (Mourad *et al.*, 2010).

Data on crop production and exports in Palestine is limited, but shows an export profile similar to the other riparian states. In addition to the same export profile as the other riparians, Syria also produces and exports apples, cotton lint, wheat, oranges and sugar beet, which is processed and later exported as refined sugar (FAOSTAT, 2016).

Crop production and water use

Mourad et al. (2010) determined that bananas and tree fruits are highly water intensive. Bananas, whilst being an economically profitable crop, especially in the international market, is highly water intensive and cannot tolerate saline water, meaning banana production and sales have led to high profits for farmers and producers in the short term, but increasing vulnerability in the long term. Jordan, Israel and Lebanon could benefit from switching crop production from bananas to lower water-intensive crops, which are also more tolerant to higher saline conditions.

All riparian states should evaluate which crops are most water intensive and which crops will be able to tolerate higher saline levels – especially Jordan, who seeks to utilize saline water for agriculture (Ghaemi et al., 2016). Currently, all riparian states produce citrus fruits and hence should investigate how to improve resilience either by altering irrigation methods or altering cropping patterns.

Virtual water in the Jordan Basin

Despite attempts and ambitions of riparian states to reduce water wastage, improve water efficiency, water availability, the production of water-intensive crops and more, their current conceptualisation of water only acknowledges physical/visible water and hence excludes virtual water from their analysis.

According to the most recent FAO statistics (FAOSTAT, 2016) riparian states of the Jordan river overwhelmingly import water-intensive crops/products such as grains and cereals; wheat, maize, soybean, rice and barley alongside potatoes (Jordan, Syria and Lebanon), sugar, both raw (Syria and Lebanon) and refined (Jordan and Lebanon), along with other goods such as oils, fruit and vegetables (Palestine), allowing states to produce less water-intensive crops such as vegetables and some fruits.

On the effects of virtual water

Virtual water is an effective tool for meeting deficits in an invisible, politically silent, and flexible manner, making it a popular choice amongst riparian states (Allan, T., 2005: 3–5). Demand for food that cannot be met locally can now be met by importing goods (i.e. virtual water), produced and sold through international markets. Hence, riparian states are not self-sufficient in water and food production; rather they are dependent on international markets and virtual water (Allan, J. A., 2002: 258). The availability of food creates a perception that the impacts of climate change and water scarcity are not as severe as in actuality or perhaps do not exist, allowing politicians to avoid controversy. Politicians, the agricultural sector, and media reinforce the sanctioned discourse, suggesting that they only need a little more water to be secure, which can be achieved using technology such as desalination or drip irrigation can (Allan, J. A., 2002: 258).

Virtual water can be mobilized quickly and be moulded to meet the changing needs of the market. For instance, if there is a crisis such as a drought or a lack of yield during the process of modifying crop patterns, riparian states can rely on the international market to meet deficits and provide food. It is flexible and fast, unlike the process of creating policies and undergoing negotiations regarding water allocation or agricultural policies, for instances.

Though virtual water seems beneficial, as it meets the demands of riparian states and does so in a silent way avoiding conflict, virtual water, neither alone nor in combination with desalination technology, presents a long-term solution for food security in the Jordan Basin. Virtual water avoids conflicts between various stakeholders due to its special characteristics, but it also prevents opportunities to address water security and food security issues, further preventing transboundary communication, cooperation, governance and water management. Though conflict is avoided, only the symptoms of water issues are temporarily treated; virtual water does not address the deeper-rooted issues of asymmetric powers, ongoing conflicts between Israel-Palestine, human rights violations and more.

Virtual water creates a dependency on imports, increasing vulnerabilities to commodity volatility, increased prices of goods and availability of the good. Riparian states substitute in-house production of crops offshore, thus increasing vulnerability to risks associated with environmental change and trade that farmers outside of riparian state borders face (Wichelns, 2001: 138). Risks include droughts, floods, lack of arable land, varying temperatures, and rainfall. Thus, if crop production is hindered abroad in a country where riparian states would normal import from, riparian states too are impacted and may not be able to import goods. When consumers realize there are insufficient goods, politicians, the agriculture sector, and media will be forced to address water issues and food security concerns.

Riparian states are not only susceptible to such risks as importers but also as exporters. If riparian states face agricultural difficulties, the economy and purchasing power of inhabitants, as well as the well-being of farmers, will be negatively affected. Further, dependence on the international market means local small and medium enterprises (SME) may miss opportunities to establish or expand a business if consumers predominantly buy from global markets, as they will be unable to compete.

Imported virtual water alongside goods produced using desalinated water is neither affordable nor accessible to all equally within and across riparian states. Desalinated water is costly to produce due to the infrastructure, energy, and other overhead costs. Thus, products created using desalinated water are likely to be sold at a higher cost. The RSDSC plan entails Israel producing water that can be sold to Palestine. However, given the poor economy in Palestine, most people will not be able to afford such water. Crops adapted to suit higher saline conditions can also be sold at a premium market price as the increased levels of salinity extend shelf-life (Mizrahi and Pasternack, 1985 in Oron et al., 2002: 237). Shaul Ben-Dov, an agronomist at Ramat Rachel expresses, 'now there is no problem of water.... The price is higher, but we can live a normal life in a country that is half desert' (Kershner, 2015: 75). Higher prices mean that only more well-off segments of communities can buy such products, creating a divide between socio-economic classes. Further, for those that can afford the increased prices of goods, water issues become less apparent and pressing. When the perception of threat is lowered, or eliminated for the rich and/or powerful, there is less advocacy for developing better policies and creating a solution to water-related issues, which may hinder progression towards reducing vulnerability for all. Such technologies and reliance on imports maintain the status quo, ensuring that the rich are better equipped to adapt to the changing nature of threats imposed by climate change, whereas the poor remain vulnerable.

Virtual water also relies on land use outside of the importing country's borders, so freeing up domestic territory for other (possibly unwise) uses. As the effects of climate change continue to impact agriculture and water, there is a likelihood that acquisition of land or 'land grabbing' will increase (Swatuk, 2018).

In the context of Israel–Palestine, Israel partakes in both land and water grabs. Israeli settlements are beside the water, furthering Israel's claims on Palestinian land established in the Armistice agreement (Malone, 2004). Ironically, Israel utilizes grabbed land and water for agriculture purposes, the produce from which is then sold back to Palestinians at a price (Gasteyer et al., 2012: 462). In this context '[w]ater grabs are rapidly adding to water scarcity for the poor in growing areas of the world', increasing vulnerability and the level of violence experienced in their everyday lives (Sassen, 2014: 191).

Cooperation and analysis

Framing water as a scarce resource may hinder cooperation and genuine long-lasting change. If water is conceptualized as scarce then policies will be created to securitize the resource (Scanlon et al. 2004: 24). Rather, water scarcity should be recognized predominantly as a consequence of poor management and governance. Diverting water, over-extracting, exploiting water sources and manipulating water allocations in the name of power, security and nation-building exemplify human involvement. Thus, accepting that water scarcity is a problem constructed by humans, it is possible to construct a solution (Swatuk, 2017). Further still, it is important to adapt our perspectives towards the environment and an entity itself that is not meant for the sole purpose of exploitation for economic gain. More so, water scarcity in the Jordan River Basin is a transboundary issue and requires a transboundary solution involving all riparian states; there is a shared responsibility. To achieve this, cooperation towards developing good governance and water-management policies amongst riparian states is necessary. But first, it is important to unravel the term 'cooperation'.

The term cooperation implies riparian states working together to achieve a common goal. It is often conceptualized as a binary opposite to conflict and hence cooperation is understood as a good thing whilst conflict is understood as bad. As Grey *et al.* (2009: 15) explain, states 'cooperate when the net benefits of cooperation are perceived to be greater than the net benefits of non-cooperation, and when the distribution of these net benefits is perceived to be fair'. The overview on water allocations, past negotiations and agreements demonstrate that not all states benefit to the same extent as the hegemon of the basin, given some states are excluded from a process of cooperation. In the Jordan Basin, the hegemon defines the common goal, and hence cooperation usually works to benefit Israel the most, and then Jordan, due to their agreements on water (Zeitoun, 2008). Thus, cooperation can reflect asymmetric power relations and in turn maintain the status quo, which in the case of the Jordan River Basin poses danger and violence. The Memorandum of Understanding signed by Israel, Jordan and the Palestinian Authority is an example of cooperation that can reinforce asymmetric power. Whilst Merekot will sell water to the Palestinian Authority (PA), there is a limitation on how much the PA can purchase. In addition, there is no guarantee that water will reach the PA in its entirety, given Israel's history of restricting water flows. Additionally, only the PA is made to purchase water, unlike Israel or Jordan who can obtain free water, failing to recognize Palestine's water rights. Thus, although Palestinians are to retrieve more water than before signing the MoU, their involvement in decision-making remains limited.

Ongoing conflicts underlie asymmetric power relations. The relationship between Israel and Palestine determines how water is managed and under what power relations. Achieving peace is a parallel process to solving the water crisis rather than a prerequisite, as the conflict involves more than just water (Jerusalem, refugees, land). Whilst manufactured water aids in alleviating quantitative problems, it does nothing to overcome asymmetric power relations or to prevent future conflicts (Bismuth, 2016: 202). As years of conflict have fostered hatred, violence and hostility between two communities, the recognition of rights and the fostering of trust and mutual understanding between communities are essential.

Friends of the Earth Middle East coordinators organize grassroots environmental education and public-awareness activities in 25 communities in Israel, Jordan and Palestine amongst the youth and adults who form a link to municipal representatives (Mehyar *et al.*, 2014: 267). One activity, 'Neighbour Path Tours', provides knowledge on the state of the Jordan River and tributaries, how waters are diverted, the effects of pollution and missed economic opportunities due to the poor health of the river (Mehyar *et al.*, 2014: 267; Jägerskog and Zeitoun, 2009: 28). Such tours provide an opportunity for participants to learn about the water issues of neighbouring communities that many often do not get, due to various reasons, including not being able to gain access to another country. The involvement of media in the tours enables those regionally and internationally to learn about the Lower Jordan River and provides a much-needed nudge to spark debates encouraging locals to voice concerns (Mehyar *et al.*, 2014: 267). Given that virtual water enables politicians to keep water issues silent, such activities foster community engagement and encourage local people to get involved and to continue raising awareness. Change that begins from the grassroots may be able to pressure governments to act. In addition, it is a positive step towards fostering a notion of caring about the source for more than the goal to securitize water for nation-state building reasons.

Cooperation is traditionally recognized in the form of agreements and nego-tiations; however, such formal recognition of cooperation also assumes onus on the state as legitimate actors forming a state-centric approach to solutions. However, cooperation should mobilize support in civil society through grass-roots initiatives working from a bottom-up approach. The responsibility of redu-cing water scarcity need not fall on the shoulders of just government but rather should be shared with all stakeholders at all levels using an Integrated Water Resource Management approach (IWRM) (Moriarty *et al.*, 2004). Granted, IWRM has its limitations, yet the aspect of involving all stakeholders at all levels remains integral. It is important to involve everyone in decision-making, as equal and acquitted stakeholders, to derive a sustainable solution. Jordan has already acknowledged the need for an Integrated Water Resources Management approach and recognizes the importance of multi-stakeholder partnerships, col-laboration and cooperation between public and private actors that can be iden-tified in the Water-DROP project (UNFCCC, 2015b: 12).

Creating synergies between riparian states, such as encouraging the transfer of knowledge, could aid riparian states to improve their water management and governance. Israel can share technological knowledge on improving drip and micro-irrigation systems and desalination with Jordan and Lebanon to help them meet their adaptation and mitigation goals. The transfer of knowledge supported by financial support from either a basin or international level can lead to improvements in irrigation techniques, increasing water availability and reducing water waste thus increasing efficiency. Jordan implemented the 'Adaptation to Climate Change to Sustain Jordan's MDG Achievements' between 2009–2013, which was a capacity-building strategy that targeted stake-holders at various levels, and created training programmes for the local com-munity including youth, decision makers and professionals. A model farm was created and used as a training and demonstration centre to show participants how to utilize reused treated water efficiently (UNFCCC, 2015b: 11). Jordan may consider inviting Lebanon to learn via training programmes and capacity-building activities to transfer knowledge that Lebanon can then integrate into its water management scheme.

Conclusion

In this chapter, we have argued that virtual water cannot provide long-term food security to the basin, though the special characteristics of virtual water (being invisible, politically silent and flexible) allow states immediate relief from food and water scarcities by importing goods and commodities to meet demands. However, it does so in a harmful way, in particular by facilitating state-centric approaches in support of the status quo. The same may be said about commitments to desalination technology, particularly in terms of Israeli commitments to climate change mitigation and adaptation plans and actions. Such approaches divide the region at a time when climate change necessitates integrated thinking, planning and practice.

At the outset, we hypothesized that by focusing on climate change activities, riparian states can desecuritize their state-centric and fragmented approaches to water and food security in the basin. After all, the Jordan River Basin is an integrated system, so long-term sustainability requires integrated approaches to governance and management. Put differently, we hoped to find evidence that by focusing on climate change and environmental cooperation, it may be possible to step away from hydro-politics to confront issues, vulnerabilities and violence that riparian states experience because of the status quo. Granted, climate change is often regarded as too abstract as a basis for motivated action, often considered an empty signifier or a buzzword, which ends up losing all meaning and weight (Methmann, 2010), but, at the outset of this project we perceived that, in this deeply divided basin, 'vagueness' might be a reasonable basis upon which to build regional cooperation. However, based on the evidence presented above, it appears to us that the highly securitized nature of all aspects of regional relations inhibit the emergence of even the most basic forms of cooperation. In relation to the water–food–climate change nexus, Israel remains committed to 'going it alone' through the application of technology and capital. This bodes ill for future regional relations and responses to climate change.

Note

1 Virtual water refers to

> water embedded in water-intensive commodities such as grain; 1,000 cubic meters of water are required to produce one cubic meter of wheat. By importing a cubic meter of wheat, a water-poor economy avoids all the economic costs and political stress of mobilizing 1,000 cubic meters of water

that can be used for other means (Allan, T, 2005: 3).

Bibliography:

Al Jazeera (2013). Al-Nakba [online]. Aljazeera.com Available at: www.aljazeera.com/programmes/specialseries/2013/05/20135612348774619.html [Accessed 15 March 2016].

Al Jazeera (2016). Rami Hamdallah: Israel waging water war on Palestinians. [online] Aljazeera.com. Available at: www.aljazeera.com/news/2016/06/israel-cuts-water-supplies-west-bank-ramadan-160614205022059.html [Accessed 12 August 2016].

Allan, J. A. (2002). Hydro-peace in the Middle East: Why no water wars?: A case study of the Jordan River Basin. *SAIS review*, 22(2), 255–272.

Allan, J. A. (2003). *Virtual water-the water, food, and trade nexus. Useful concept or misleading metaphor?*. Water international, 28(1), pp. 106–113.

Allan, T. (2005). 'Virtual Water': a long term solution for water short Middle Eastern economies? Water Issues Group, School of Oriental and African Studies, University of London.

Bismuth, C. (2016). Water resources, cooperation and power asymmetries in the water management of the lower jordan valley: the situation today and the path that has led there. *Society-Water-Technology*, 189–204.

Dreizin, Y., Tenne, A., and Hoffman, D. (2007). Integrating large scale seawater desalination plants within Israel's water supply system. *Desalination*, 220(1), 132–149.

Einav, R., Harussi, K. and Perry, D., 2002. The footprint of the desalination processes on the environment. *Desalination*, 152(1), pp. 141–154.

Einav, R. and Lokiec, F., 2003. Environmental aspects of a desalination plant in Ashkelon. *Desalination*, 156(1), pp. 79–85.

FAO Aquastat (2009). Jordan Basin. Available at: www.fao.org/nr/water/aquastat/basins/jordan/index.stm [Accessed 16 May 2018].

FAOSTAT (2016). Commodities. [online] FAOSTAT. Available at: www.fao.org/faostat/en/#rankings/commodities_by_country [Accessed 30 March. 2018].

Gasteyer, S., Isaac, J., Hillal, J., and Walsh, S. (2012). Water grabbing in colonial perspective: Land and water in Israel/Palestine. *Water Alternatives*, 5(2), 450–468.

Ghaemi, A. A., Dindarlou, A., Golmakani, M. T., and Razzaghi, F. (2016). The effect of salinity and irrigation regimes on the level of fatty acids in olive flesh oil. *Modern Applied Science*, 10(5), 98–111.

Gleick, P. H. (2004). *The World's Water 2004–2005: The Biennial Report on Freshwater Resources*. Washington DC: Island Press.

Grey, D., Sadoff, C., and Connors, G. (2009). Effective cooperation on transboundary waters: A practical perspective. *Getting Transboundary Water Right: Theory and Practice for Effective Cooperation, Report*, (25), 15–20.

Habib, J. (2007). Both sides now: Reflections on the Israel/Palestine conflict. *Human Rights Quarterly*, 29(4), 1098–1118.

IMFA (2016). The Balfour Declaration [online]. Available at: www.mfa.gov.il/MFA/ForeignPolicy/Peace/Guide/Pages/The%20Balfour%20Declaration.aspx [Accessed 11 April 2016].

IRIN (2018). Palestine refugees: locations and numbers. Available at: www.irinnews.org/report/89571/middle-east-palestinian-refugee-numberswhereabouts [Accessed 30 March 2018].

Israel Ministry of Environmental Protection (IMEP) (2009). Coping with climate change in Israel. Special Edition [pdf]. Israel Ministry of Environmental Protection. Available at: www.sviva.gov.il/English/env_topics/climatechange/Documents/CopingWithClimate ChangeInIsrael-SpecialEBulletin-Dec2009.pdf [Accessed 18 February 2016].

Israel Ministry of Environmental Protection (IMEP) (2016). Targeting climate change in Israel: toward Paris and beyond. 42nd edn [pdf] Israel environment bulletin, pp. 8–16. Available at: www.sviva.gov.il/English/env_topics/climatechange/Mitigation/Documents/Targeting%20Climate%20Change%20in%20Israel%20Toward%20

Paris%20and%20Beyond%20-%20Israel%20Environment%20Bulletin%20-%20 Jan%202016.pdf [Accessed 18 February 2016].

Jägerskog, A., and Zeitoun, M. (2009). *Getting Transboundary Water Right: Theory and Practice for Effective Cooperation*. Stockholm: Stockholm International Water Institute.

Kally, E., and Fishelson, G. (1993). *Water andPeace: Water Resources and the Arab-Israeli Peace Process*. New York: Praeger.

Kershner, I. (2015). Aided by the sea, Israel overcomes an old foe: Drought. *New York Times* (29 May).

Luft, G. (2002). The Wazzani Water Dispute: More tension along the Israel–Lebanon Border. [online] The Washinton Institute. Available at: www.washingtoninstitute.org/ policy-analysis/view/the-wazzani-water-dispute-more-tension-along-the-israel-lebanon-border [Accessed 2 December 2016]

Luo, T., Young, R., and Reig, P. (2015). Aqueduct projected water stress country rankings. [online] www.wri.org/sites/default/files/aqueduct-water-stress-country-rankings-technical-note.pdf. Available at: www.wri.org/sites/default/files/aqueduct-water-stress-country-rankings-technical-note.pdf [Accessed 6 March 2016].

Malone, A. R. (2004). Water now: the impact of Israel's security fence on Palestinian water rights and agriculture in the West Bank. *Case Western Reserve Journal of International Law*, 36(2), 639–671.

Mehyar, M., Al Khateeb, N., Bromberg, G., and Koch-Ya'ari, E. (2014). Transboundary cooperation in the Lower Jordan River Basin. In: E. Weinthal, J. Troell and M. Nakayam, eds, *Water and Post-Conflict Peacebuilding*. London: Earthscan.

Meisen, P., and Tatum, J. (2011). The water-energy nexus in the Jordan River Basin: The potential for building peace through sustainability. Global Energy Network Institute. Available at:*www.geni.org/globalenergy/research/water-energy-nexus-in-the-jordan-river-basin/the-jordan-river-basin-final-report.pdf* [Accessed 20 February 2016].

Methmann, C. P. (2010). 'Climate protection' as empty signifier: A discourse theoretical perspective on climate mainstreaming in world politics. *Millennium-Journal of International Studies*, 39(2), 345–372.

Moriarty, P., Butterworth, J., and Batchelor, C. (2004). Integrated water resources management and the domestic water and sanitation sub-sector. IRC Thematic Overview Paper, IRC International Water and Sanitation Centre, Delft, The Netherlands.

Mourad, K. A., Gaese, H., and Jabarin, A. S. (2010). Economic value of tree fruit production in Jordan Valley from a virtual water perspective. Water *Resources Management*, 24(10), 2021–2034.

Netafim (2015). Glossary – Netafim. [online] Netafim.com. Available at: www.netafim. com/glossary#d [Accessed 23 September 2016].

Oron, G., DeMalach, Y., Gillerman, L., David, I., and Lurie, S. (2002). SW – soil and water: effect of water salinity and irrigation technology on yield and quality of pears. *Biosystems Engineering*, 81(2), 237–247.

Sassen, S. (2014). *Expulsions: Brutality and Complexity in the Global Economy*. Cambridge, MA: Belknap Press.

Scanlon, J., Cassar, A., and Nemes, N. (2004). Water as a human right? (51). IUCN.

Swatuk, L. A. (2018). The land-water-food-energy nexus: green and blue water dynamics in contemporary Africa-Asia relations. In: P. M. A. Raposo de Medeiros Carvalho, D. Arase and S. Cornelissen, eds, *Routledge Handbook of Africa-Asia Relations*. New York and London: Routledge.

Swatuk, L. A. (2017). *Water in Southern Africa*. Pietermaritzburg: UKZN Press.

Treaty of Peace (1994). Treaty of Peace Between the State of Israel and the Hashemite Kingdom of Jordan 1994, 26 October 1994. Available at: www.mfa.gov.il/MFA/ForeignPolicy/Peace/Guide/Pages/Israel-Jordan%20Peace%20Treaty%20Annex%20II.aspx [Accessed 20 April 2016].

UNFCCC (2015a). Israel NDC. Available at: www4.unfccc.int/submissions/INDC/Published%20Documents/Israel/1/Israel%20INDC.pdf [Accessed 11 March 2016].

UNFCCC (2015b). Jordan NDC. Available at: www4.unfccc.int/submissions/INDC/Published%20Documents/Jordan/1/Jordan%20INDCs%20Final.pdf [Accessed 11 March 2016].

UNFCCC (2015c). Lebanon NDC. Available at: www4.unfccc.int/submissions/INDC/Published%20Documents/Lebanon/1/Republic%20of%20Lebanon%20-%20INDC%20-%20September%202015.pdf [Accessed 11 March 2016].

UNFCCC (2016). Paris. Available at: http://unfccc.int/paris_agreement/items/9485.php

UNHCR (2018). Syria regional refugee response. Last data update 22 March 2018. https://data2.unhcr.org/en/situations/syria [Accessed 30 April 2018].

Weinthal, E., Zawahri, N., and Sowers, J. (2015). Securitizing water, climate, and migration in Israel, Jordan, and Syria. *International Environmental Agreements: Politics, Law and Economics*, 15(3), 293–307.

Wichelns, D. (2001). The role of 'virtual water' in efforts to achieve food security and other national goals, with an example from Egypt. *Agricultural Water Management*, 49(2), 131–151.

World Bank (2013a). Statement on water sharing agreement signed by Israeli, Jordanian and Palestinian representatives. Available at: www.worldbank.org/en/news/press-release/2013/12/18/clarification-water-sharing-agreement-israeli-jordanian-palestinian-representatives [Accessed 12 March 2016].

World Bank (2013b). Senior Israeli, Jordanian and Palestinian representatives sign milestone water sharing agreement. Available at: www.worldbank.org/en/news/press-release/2013/12/09/senior-israel-jordanian-palestinian-representatives-water-sharing-agreement [Accessed 11 March 2016].

World Bank (2016a). Climate wizard output. [online] Climatewizard.ciat.cgiar.org. Available at: http://climatewizard.ciat.cgiar.org/outputs/JordanRiverBasin/ [Accessed 10 November 2016].

World Bank (2016b). Climate change knowledge portal 2.0. [online] Sdwebx.worldbank.org. Available at: http://sdwebx.worldbank.org/climateportal/index.cfm?page=country_future_climateandThisRegion=Middle [Accessed 13 March 2016].

World Bank (2016c). Climate-driven water scarcity could hit economic growth by up to 6 percent in some regions, says World Bank [press release]. Available from: www.worldbank.org/en/news/press-release/2016/05/03/climate-driven-water-scarcity-could-hit-economic-growth-by-up-to-6-percent-in-some-regions-says-world-bank. [Accessed 3 March 2016].

World Economic Forum (WEF) (2016). *The Global Risks Report 2016*. Geneva: World Economic Forum.

World Economic Forum (WEF) (2018). *The Global Risks Report 2018*. Geneva: World Economic Forum.

Yeung, P. (2016). Israel 'cuts off water supply to West Bank' during Ramadan. [online] *Independent*. Available at: www.independent.co.uk/news/world/middle-east/ramadan-2016-israel-water-west-bank-cuts-off-a7082826.html [Accessed 12 Aug. 2016].

Zeitoun, M. (2008). *Power and Water in the Middle East: The Hidden Politics of the Palestinian-Israeli Water Conflict*. I. B. Tauris.

9 A gendered analysis of Integrated Soil Fertility Management (ISFM) as a strategy for strengthening adaptive capacity in Ghana's Tolon District

Alhassan Lansah Abdulai and Sebastiaan Soeters

Introduction

The Ghana National Climate Change Adaption Strategy (RoG, 2016) has highlighted climate change as a potential trigger for low productivity in agriculture, for reducing both quantity and quality of water, for unsustainable harvesting of natural resources, and for increased incidences of water, air, and food borne diseases (MESTI, 2013). A number of strategies have emerged that seek, sometimes implicitly, to strengthen the adaptive capacity and resilience of farming communities in northern Ghana by, in one way or the other, increasing agricultural productivity. For instance, building dams and dugouts under the Government of Ghana's (GoG) 'One Village, One Dam' policy seeks to make farming a year-round enterprise and thereby diversify income, and make farming less dependent on the increasingly unpredictable rainfall pattern. Furthermore, Sanchez and Jama (2002) attribute the inability to match food supply to demand in sub-Saharan Africa to soil nutrient depletion resulting from the intensification of land use without proper land-management practices and inadequate external inputs. According to Chauvin *et al.* (2012), declining per capita food production couples with the rapidly growing populations (estimated at 3 per cent per annum) to reduce per capita food availability for households of smallholder subsistence farmers. As a result, addressing soil fertility has emerged as another important pillar of northern Ghana's emerging adaptation regime, promoted by the GoG and donors alike. In this regard, a plethora of views, paradigms, and concepts have arisen related to soil fertility management. These include integrated natural resource management (INRM), integrated nutrient management (INM), system of rice intensification (SRI), conservation agriculture (CA), organic agriculture (OA), integrated pest management (IPM), agroforestry (AF), precision agriculture (PA) and many others (Rosegrant *et al.*, 2014; Lee, 2005).

There is by now a well-developed 'scientistic' literature on methods of soil conservation. It is clear that Integrated soil fertility management (ISFM) proves to be an effective means of adapting to climatic change and variability in sub-Saharan Africa (SSA), where low and declining soil fertility is a major cause of low productivity of the predominantly smallholder and rain-fed agricultural systems, and the consequent persistent poverty and food insecurity. However, whilst the scientific community unanimously regards ISFM as a scientifically proven strategy, and invests resources into its development, evaluation and promotion, very little knowledge exists on the gender dimensions of these processes and its potential for conflict. This chapter therefore assesses the unintended impacts of ISFM practices on gender relations and conflict in the interior Savannah of Ghana. It focuses on communities where Council for Scientific and Industrial Research, through the Savanna Agricultural Research Institute (CSIR-SARI). has been involved in the implementation of ISFM as a strategy for both increasing crop productivity and as an adaptation to climate change.

Definition, concept and rational of ISFM and conceptualizing its relations to exclusion

Vanlauwe *et al.* (2010) define ISFM as

> a set of soil fertility management practices that necessarily include the use of improved germplasm, chemical fertilizer, and organic inputs, combined with the knowledge on how to adapt these practices to local conditions, aiming to maximize agronomic use efficiency of the applied nutrients and improve crop productivity.

ISFM involves an integrated application of a set of good soil-management technologies with the aim of exploiting complementarities and synergies among the different technologies that individually can have positive contributions to soil fertility, agronomic efficiency and crop productivity (Vanlauwe *et al.*, 2011; Vanlauwe *et al.*, 2010; Marenya and Barrett, 2007; Place *et al.*, 2003). Improvement of agronomic efficiency is central to ISFM, and Figure 9.1 shows a conceptual relationship between advancement towards full ISFM and agronomic efficiency.

The fundamentals of ISFM are that agricultural intensification cannot occur without investments in soil health, and that we require both organic and mineral inputs to sustain soil health and increase crop production (Vanlauwe *et al.*, 2010). The underlining principles of ISFM include:

1   Well-adapted stress tolerant genotypes or cultivars are necessary for efficient use of available nutrients
2   The combined use of mineral fertilizers and organic matter is critical since the sole use of each of them is not sufficient for sustainable agricultural production
3   Good agronomic practices (planting dates, planting densities, and weed control, disease and insect pest control, timely harvesting) are essential for ensuring the efficient use of scarce nutrient and other resources.

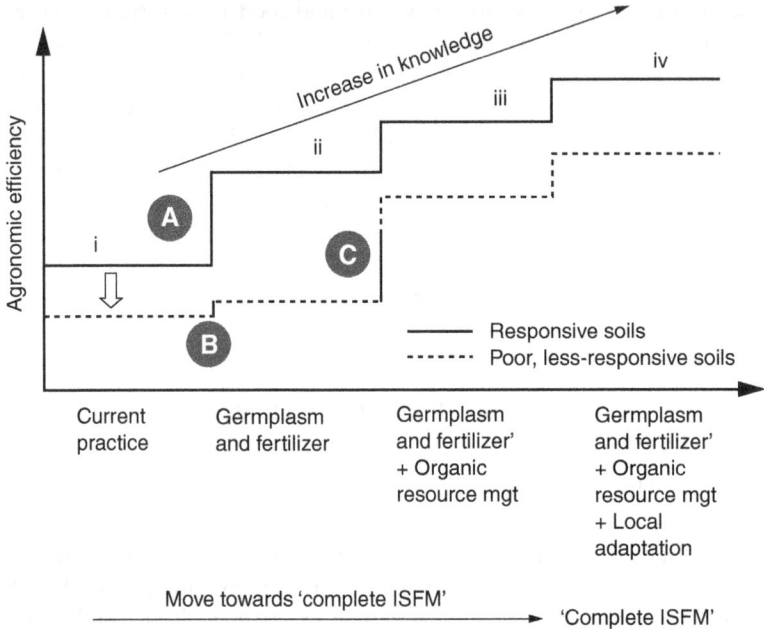

Figure 9.1 Conceptual relationship between the agronomic efficiency and the implementation of various components of Integrated Soil Fertility Management (ISFM).

In addition to these principles, ISFM recognizes the need to target nutrient
resources within intercropping and crop rotation cycles, preferably including
legumes, thus going beyond recommendations for single crops.

The theory of change for the more than $160 M Soil Health Program (SHP),
implemented by the Alliance for Green Revolution in Africa (AGRA) with
funds from the Bill and Melinda Gates and the Rockefeller Foundations
((BMGF and RF), demonstrates the importance of ISFM in African agriculture.
It states: 'To increase their soil productivity and achieve sustainable yield
increases, smallholder farmers need to adopt and use integrated soil fertility
management practices in an economically viable way' (AGRA and IIRR, 2014).
SHP was a core component of AGRA's efforts to rejuvenate the highly degraded
and heavily depleted soils of Africa through the deployment of ISFM and
increased access to fertilizer in 13 African countries.

Savanna Agricultural Research Institute (SARI), one of the 13 autonomous
institutes under the Council for Scientific and Industrial Research (CSIR), and
known in short as CSIR-SARI, has been a major collaborator in the promotion
of Integrated Soil Fertility Management (ISFM) in Ghana. ISFM is important
for CSIR-SARI for a number of reasons. First, the major goal of the institute is
to improve the livelihoods of rural households through the development, identi-
fication, introduction, evaluation, promotion and dissemination of appropriate

agricultural technologies to enhance the productivity of cereal-, legume-, and root and tuber-based cropping systems in the interior Savannah zone of Ghana. Second, agricultural production systems in the mandate area of CSIR-SARI are highly sensitive and vulnerable because soils in this area are generally poor in fertility, heavily eroded, shallow, have low moisture holding capacities and/or infiltration rates and are prone to run-off. Third, both agricultural productivity and production in the interior Savannah zone of Ghana are low because irrigation coverage is very low (less than 5 per cent), so food and cash crop cultivation is mainly rain-fed and therefore negatively affected by the synergy of low soil fertility and impacts of climate change. Finally, ISFM does not only improve soil fertility, but can double as a strategy for climate change adaptation.

Lines of gender exclusion

Since climate change affects the most vulnerable groups most profoundly, we need to assess and select interventions that aim to strengthen adaptive capacity based on their ability to influence the livelihoods of those groups positively. As a result, understanding how ISFM affects women's livelihoods (since women ordinarily represent the most vulnerable community members) serves as a legitimate way of evaluating its success. Importantly, making sense of who is included in efforts to improve soil fertility, and who is not, depend on a number of lines of exclusion. This refers to conditions necessary for inclusion in the benefits of ISFM. More specifically, in order for ISFM to, for instance, work as a driver of strengthening adaptive capacity, will depend upon access to manure, access to land, access to improved cultivars and finally, the appropriateness of crops beings cultivated. Whilst ISFM appears in 'scientistic' discourses, as a remedy for decreasing yields, there is no equitable distribution and access of the resources or conditions required for reaping its benefits, with strong socio-economic characteristics and livelihoods relating to gender driving various forms of access. These barriers to access form the basis of our analysis. In short, we attempt to understand gender dynamics of each of these barriers and, based on access, draw conclusions about the impact of ISFM on men and women. How much benefit you derive from ISFM depends on access to these resources, with the benefits increasing with number of different resources one can access. Figure 9.2 shows the differentials in access to the resources between males and females in the study area.

Description of the study area

The study was conducted in the Tolon District, located between latitudes 9°15' and 10°02' North and longitudes 0°53' and 1°25' West. It shares boundaries to the North with Kumbungu, North Gonja to the West, Central Gonja to the South and Sagnarigu Districts to the East. The district experiences a mono-modal rainfall season, which starts in late April (with little rainfall), rises to its peak from July to August and declines sharply to a complete halt from October

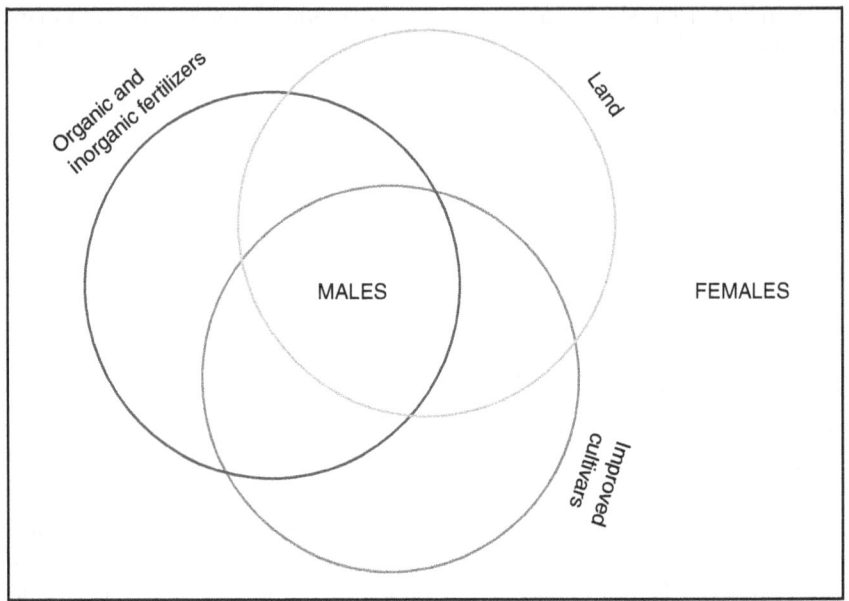

Figure 9.2 Conditions for the inclusion in benefits of ISFM.

to November. The dry season runs from November to March, with day temperatures ranging from 33°C to 39°C, while mean night temperatures range from 20°C to 26°C. The mean annual rainfall ranges between 950 mm and 1,200 mm. The area experiences occasional storms, which have implications for bare soil erosion depending on its frequency and intensity, especially when they occur at the end of the dry season. The situation also influences staple crop farming that is already highly restricted by the short rainfall duration.

The main vegetation is grassland, interspersed with guinea savannah woodland, characterized by drought-resistant trees such as acacia, (*Acacia longifolia*), mango (*Mangiferaindica*), baobab (*Adansoniadigitata* Linn), shea nut (*Vitellaria-paradoxa*), dawadawa, (*Parkiabiglobosa*) and neem (*Azadirachtaindica*). Major economic tree species include the sheanut, dawadawa and mango, which form an integral part of the livelihood of the people. There is also the neem, which is mostly used for medicinal purposes. One undesirable characteristic feature of the district is the annual bush fires, which sweep across the savannah woodland and exacerbate the impact of both wind and water erosion.

According to the housing and population census of 2010, about 92.4 per cent of the population of Tolon District engages in agriculture, with crop farming being the main agricultural activity. Those in livestock rearing account for 74.1 per cent and tree planting 0.7 per cent. In the rural localities, more than nine out of every ten (96.6%) of the households are agricultural households and 65.4 per cent are in the urban localities. Poultry (chicken – 36.8 per cent) is the dominant animal reared in the district. The success of agriculture, the mainstay of the people of the Tolon district, depends largely on the quality of the soil.

Selection of communities

The project team selected three communities within the Tolon district where CSIR-SARI has been promoting ISFM to assess the impacts of the intervention on gender dimensions with regard to access to opportunities and/or resources to increase productivity and production by enhancing agronomic efficiency. The project team selected three communities based on their association with the demonstration of interventions developed by CSIR-SARI. These communities were Wantugu (9°30'48.29'N; 1°08'52.23'W), Dimabi (9°24'36.86'N; 1°05'03.92'W) and Kpalsogu (9°24'12.49'N; 1°00'16.39'W), located within the Tolon district of the Northern region of Ghana (Figure 9.3).

Each of the communities selected for this study has had one or more soil-fertility interventions. CSIR-SARI and the Ministry of Food and Agriculture (MOFA) implemented the Soil Health Program (SHP) at Kpalsogu. At Dimabi,

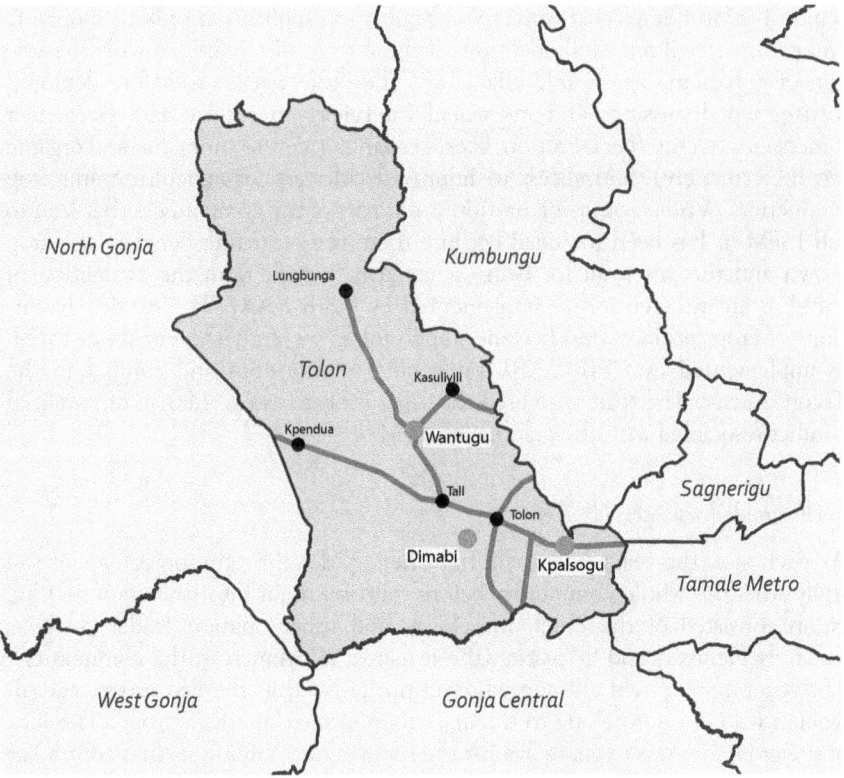

Figure 9.3 Map of Tolon District showing the locations of the study.

ISFM was deployed under a project funded by World Bank through CORAF/ WECARD and entitled 'Enhancing Resilience and Adaptive Capacity to Climate Change through Integrated Land, Water and Nutrient Management in Semi-Arid West Africa (ENRACCA-WA)', as an intervention for enhancing the resilience and adaptive capacity to climate change of maize farmers by CSIR-SARI. Finally, at Wantugu, demonstrations on ISFM were carried out by the Ministry of Food and Agriculture (MoFA) and CSIR-SARI (then Nyank-pala Agricultural Experimental Station; NAES).

Methodology

The research team entered the selected communities in January 2014 using the agricultural extension agents (AEAs) of the department of agriculture (DOA) responsible for these communities as a conduit. At such meetings, the team explained the objectives of the study to an assembly comprised of the chief and elders, opinion leaders, and members of each community. At these meetings, dates were set for detailed discussions on the project as well as profiling of communities regarding their contact with and practice of ISFM.

The conceptual framework for moving to full ISFM consists of using responsive land, improved genotypes or cultivars, application of both inorganic and organic fertilizers, and adapting all these to local conditions with the sole aim of enhancing agronomic efficiency. The project team therefore deployed focus-group discussions and individual interviews to collect data on gender dimensions of crop specialization, access to land, access to inorganic and organic fertilizer (manure), and access to improved cultivars for agriculture and crop production. While crop specialization is not part of the components that lead to full ISFM, it has been included because there is a clear link between the crop grown and the potential for being selected to benefit from the evaluation of ISFM as an intervention as implemented by CSIR-SARI. The gender dimensions of crop specialization become important as we study the impact of ISFM, as implemented by CSIR-SARI, on gender relationships and conflict in the Tolon district. The team also used the same tools to assess the risk of potential conflict associated with the practice of ISFM.

Focus-group discussions (FGD)

At each site, the team had an initial briefing, detailing the objectives of the study with the whole community before splitting them into three groups. One group consisted of the chief, his elders, and other opinion leaders such as assembly members and 'Magazia' (the leader of all women in the community). The second group was all men who did not fall within the first group, and all women who did not belong to the first group formed the third group. The idea of the groupings was to guard against the suppression of opinions that contradict those of the opinion leaders or suppression of opinions of females that contra-dict those of males. The discussions focused on gender dimensions of access to improved cultivars, access to inorganic and organic (manure) fertilizers, and access to land (both acreage and quality), since all these affected the successful implementation of ISFM.

Individual interviews to validate outcomes of FGD

The team also conducted a survey by interviewing individuals based on a structured questionnaire to validate the outcomes of the focus-group discussions. Details of the number of people sampled, disaggregated by community and gender are presented in Table 9.1.

Results and discussion

Crop specialization

Crops respond differentially to marginal or degraded soils. Generally, crops such as maize, rice, sorghum, millet and yam, are less tolerant of marginal soils and are therefore favourites for ISFM because of their status as major staples. Crops such as okra and amaranths tolerate marginal soils because they give reasonable yields with little or no external fertilizer input, and are therefore the least favoured when selecting candidates for ISFM. According to participants of FGDs, there is a gender specialization concerning the type of crops grown. Males grow staple crops and all those that require fertile soils or external fertilizer inputs. These crops include maize, rice, sorghum, millet, yam, cassava, garden eggs, sweet potato and cotton. On the other hand, females concentrated on crops that can tolerate marginal soils, and therefore require little or no external fertilizer input. Crops such as okra, amaranths and hibiscus or roselle, constitute the crops grown by females only. These are vegetables used in preparing soup, sauce and stew, a main duty of females. However, there were crops that both males and females cultivated because they double as crops that demand low input, and cash crops. The crops under this category are cowpea, groundnuts, soybean, pepper and tomatoes.

Women at Kpalsogu who participated in the focus-group discussion indicated that the ISFM promotion team did not select any female for direct participation and benefit because they tested the strategy on maize (a solely male crop), though some of their husbands participated. Sanatu Fuseini said, 'we do not

Table 9.1 Number of individuals interviewed disaggregated by community and gender

Community	Males	Females	Total
Dimabi	18	12	30
Kpalsogu	20	10	30
Wantugu	14	16	30
Total	52	38	90

grow maize because all our soils are bad so we have to apply fertilizer to get some yield, but we cannot afford it because fertilizer is very expensive'. Mariama Yakubu added that maize is a staple, 'so if we produce it, we cannot sell it because maize is the main grain that we use for food so our husbands own that'.

The women at Dimabi indicated that promotion of ISFM did not capture their interest because the test crop (maize) is a male-dominated crop while the use of organic matter such as manure and crop residue robbed them of a valuable household resource (biofuel), which they have enjoyed for a long time now. At Wantugu, the group of women also indicated that the team promoting ISFM did not invite any females alongside the males to observe the demonstrations on ISFM because farming then was solely a male thing. According to the women, the situation has changed because women now cultivated crops, but maize remains a males' crop because it needs good land and fertilizer that they do not have and cannot afford.

Access to land

Having access to land is the basis of benefitting from ISFM. All the males, constituting 58 per cent of the respondents, indicated that it was easy to access land; a majority of the males, 41 out of 52 (48 per cent of the respondents), were of the view that it is extremely easy to access land. On the other hand, all the female respondents indicated that it was difficult to access land, with the majority of them, 31 out of 38 (34 per cent of respondents), claiming that it was extremely difficult, especially access to fertile land (Table 9.2). These show that access to land skews towards males, thereby widening the gender gap in terms of access to land as both a natural resource and a factor of agricultural production. Both genders cited cultural reasons for the situation, and that females need male mediators (brothers or husbands) to access land from members of different families or clans. One major concern raised regarding access to land for ISFM is the fact that landowners will usually ask for their land after a year of impressive yields because of ISFM practiced by the tenant. This therefore does not allow the establishment of well-buffered agriculturally productive systems, making it a disincentive to adopt ISFM on borrowed land, hence the continuous mining of soils and degradation of land.

Table 9.2 Responses of respondents on ease of access to land for crop production

Access to land	Males	Females	Total	Percentage (%)
Extremely easy	41	0	41	46
Easy	11	0	11	12
Difficult	0	0	0	8
Extremely difficult	0	7	7	34
Not possible	0	31	31	0
Total	52	38	90	100

Ease of access to manure for crop production

Organic matter is a very important component of ISFM, especially in the study area where there are highly eroded and degraded soils with low fertility. The Soil Health Program (SHP) introduced a commercially produced organic fertilizer (Fertisol), but community members indicated that the costs of both the product and transportation are prohibitive. They also explained that crop rotation and intercropping might not be able to build the highly eroded and degraded soils because household and community members mostly clear crop residue from plots for other competitive uses such as fuel wood and animal feed. The problem they associated with compost was the inadequate amounts of material in the dry season and the lack of sufficient labour, during the wet season, due to competitive demands for labour by other farming activities. The consensus at all the focus-group discussions was that using manure is the most practicable among the options, though with some few associated problems. Discussants from all the groups indicated that both physical and economic access to manure were skewed in favour of males, since females have no part in manure collected from animal pens and also lack the economic power to buy, even if manure is available for sale.

When we conducted individual interviews to validate the findings on access to manure from the focus-group discussions, 38 per cent of the respondents (all males) said that they had easy access to manure. The rest of the respondents indicated that it was difficult (28 per cent), extremely difficult (28 per cent), or not possible (5 per cent). Out of the 38 female respondents, 37 indicated that it was either difficult or extremely difficult to access manure, while one respondent said it was not possible to access manure (Table 9.3). The main reason adduced was that males build the shelters for ruminants and therefore collect and use all the manure.

The women lamented that the test crop and their lack of access to manure and other forms of organic matter automatically disqualified them from participating in the evaluation of ISFM.

Access to improved cultivars

During the focus-group discussions, members of all groups in all the communities expressed the need for improved crop cultivars that can withstand

Table 9.3 Responses of respondents on ease of access to manure for crop production

Ease of Access to manure	Males	Females	Total	Percentage (%)
Extremely easy	0	0	0	0
Easy	35	0	35	39
Difficult	11	14	25	28
Extremely difficult	2	23	25	28
Not possible	4	1	5	5
Total	52	38	90	100

drought, heat and flood as well as noxious weeds, diseases and insect pests because their traditional cultivars cannot tolerate these stresses. They indicated that abrasive weather, disease and insect-pest outbreaks reduce crop yields when they occur. However, the few community members who participated in the evaluation of ISFM as an adaptation intervention indicated that they had access to seeds of maize cultivars and/or striga (a parasitic plant), which are tolerant of drought, but in insufficient quantities. All the beneficiaries who accessed seeds of improved cultivars were males. There were no female beneficiaries present at any of the focus-group discussions. The individual questionnaire therefore tried to capture the ease with which one who does not participate in the evaluation of ISFM as an adaptation intervention can access improved cultivars. Table 9.4 shows the respondents' opinions about access to improved cultivars of maize by non-beneficiaries of the ISFM package. All the responses ranged from 'difficult', through 'extremely difficult' to 'not possible', when we asked how easy it was for non-beneficiaries to access improved cultivars of maize. A majority of the males, (44 out of the 52) representing 49 per cent of all the respondents, indicated that it was very difficult to access improved maize cultivars tested under ISFM if one did not participate in the evaluation of the intervention.

The remaining male-respondents (8) opined that it was extremely difficult for non-beneficiaries to access the improved cultivars of maize. Almost two thirds of the female respondents (25 out of 38), representing 28 per cent of all respondents, were of the opinion that it was not possible for non-beneficiaries to access improved maize cultivars, with the remaining 13 female respondents indicating that it was extremely difficult for non-beneficiaries to access improved cultivars of maize. In total, 23 per cent of the respondents (21 respondents) indicated that accessing improved cultivars of maize was extremely difficult for non-beneficiaries of ISFM.

Risk of domestic conflict associated with ISFM

All groups in all the communities highlighted the alternative uses for crop residue during the focus-group discussions. They indicated that residue from maize, sorghum, millet and pigeon pea were used as animal feed, biofuel and for other income-generating activities such as mat- and basket-weaving.

Table 9.4 Responses on the ease of access to improved cultivars by non-beneficiaries of ISFM

Non-participant's access to improved cultivars	Males	Females	Total	Percentage (%)
Extremely easy	0	0	0	0
Easy	0	0	0	0
Difficult	44	0	44	49
Extremely difficult	8	13	21	23
Not possible	0	25	25	28
Total	52	38	90	100

Focus-group discussants also stated that of all the possible strategies of ISFM, the incorporation of crop residue had a high potential risk of triggering domestic conflict. All the respondents of the individual survey agreed that there was a risk of domestic conflict associated with the requirements and manner in which ISFM was evaluated, with 82 per cent of them opining that the risk was very high (Table 9.5). According to most of the respondents, the risk of domestic conflict resides in the redistribution of crop residue as a resource because of its competitive alternative uses by different household members and livelihood options.

Risk of intra community conflict associated with ISFM

Focus-group discussants agreed that intra-community conflicts will arise if the residue retention component of ISFM practice is adhered to. They indicated that the conflicts would arise if women from other households collected crop residue for use as biofuel, if stray animals for other community members browsed on fields where crop residue have been retained, and weavers of mats and baskets from other households collecting the crop residue retained on the field. The women groups in particular stated that it would require strenuous efforts to stop them from collecting maize and sorghum stalks retained on fields because they have no alternative sources of biofuel. The responses of 90 individuals interviewed regarding the risk of intra-community conflict associated with the adoption of ISFM are presented in Table 9.6. The proportion of respondents

Table 9.5 Responses of participants on the risk of domestic conflict associated with ISFM

Risk of domestic conflict	Males	Females	Total	Percentage (%)
Very high	42	32	74	82
High	10	6	16	18
Low	0	0	0	0
Very low	0	0	0	0
No risk	0	0	0	0
Total	52	38	90	100

Table 9.6 Responses of participants on the risk of intra-community conflict associated with ISFM

Risk of intra-community conflict	Males	Females	Total	Percentage (%)
Very high	1	27	28	31
High	44	8	52	58
Low	7	3	10	11
Very low	0	0	0	0
No risk	0	0	0	0
Total	52	38	90	100

who stated that adoption of ISFM will pose a low risk of intra-community conflict was 11 per cent, while the respective proportions of respondents stating that adoption of ISFM will pose high and very high risk of intra-community conflict were 58 per cent and 31 per cent.

Risk of inter-community conflict associated with ISFM

A majority of the discussants agreed that the potential risk of inter-community conflict associated with the adoption and practice of ISFM is low. However, some few discussants showed dissent for the majority stand and suggested that the risk of inter-community conflict is high. Indiscriminate burning of bushes by humans and uncontrolled browsing by ruminants from nearby communities who are not beneficiaries of ISFM could trigger inter-community conflict. The opinions of males and females on the potential risk of inter-community conflict associated with the adoption and practice of ISFM seemed to have converged. To validate the information gathered from the focus-group discussion on the issue, a question for the individual survey targeted that issue. Table 9.7 presents the responses of the sampled respondents on that question. None of the respondents stated that there was no risk of inter-community conflict or the risk of inter-community conflict was very high. A majority of the respondents (72 per cent) stated that the risk of inter-community conflict was low while 32 per cent was of the opinion that the risk of inter-community conflict was very low. In total, 95 per cent of the respondents were of the opinion that risk of inter-community conflict associated with the adoption and practice of ISFM is low. The outcomes of focus-group discussions and individual surveys converged, and can thus be recommended for further action on the subject of inter-community conflicts.

Conclusions and recommendation

The study has shown that the selection of beneficiaries of the ISFM as an intervention for adapting to climate change was not equitable in terms of gender. The fact that all the beneficiaries were male widened the gap between males and females in terms of access to opportunities and resources. The gender bias,

Table 9.7 Responses of participants on the risk of inter-community conflict associated with ISFM

Risk of inter-community conflict	Males	Females	Total	Percentage (%)
Very high	0	0	0	0
High	2	2	4	5
Low	42	23	65	72
Very low	8	13	21	23
No risk	0	0	0	0
Total	52	38	90	100

in favour of males, in the access to land, manure and improved seeds will syner-gize to widen the already large development and livelihood gaps between men and women regarding access to opportunities and resources. The study also brought to the fore the fact that crop residue constitutes a resource with com-petitive alternative uses and could be a source of hierarchical conflicts if issues regarding its distribution and use are not planned in consultation with all the relevant stakeholders. Finally, there are a number of complementary activities that might work to reduce the exclusionary tendency of ISFM. For instance, community members can establish woodlots to provide a source for the fuel-wood needs of women, and thereby ensure an uncontested access to crop residue for incorporation. Livestock production systems might also be employed as a means of transforming crop residue into manure, which is much more refined and can rapidly increase soil health, as well as reduce the risk of conflict due to minimized competition for crop residue.

Acknowledgements

This study has been funded by Department for International Development of the United Kingdom (DFID) through the Dutch Organisation for Scientific Research (NWO).

References

AGRA (Alliance for a Green Revolution in Africa) and IIRR (International Institute of Rural Reconstruction) (2014). Investing in soil: Cases and lessons from AGRA's Soil Health Programme. Nairobi: Alliance for a Green Revolution in Africa and International Institute of Rural Reconstruction.

Chauvin, N. D., Mulangu, F., and Porto, G. (2012). Food production and consumption trends in Sub-Saharan Africa: Prospects for the transformation of the agricultural Sector. *UNDP Working Paper. WP 2012–2011* (February), New York: UNDP.

Lee, D. R. (2005). Agricultural sustainability and technology adoption: Issues and policies for developing Countries. *American Journal of Agricultural Economics*, 87, 1325–1334.

Marenya, P., and Barrett, C. B (2007). Household-level determinants of adoption of improved natural resources management practices among smallholder farmers in western Kenya. *Food Policy*, 32, 515–536.

MESTI (Ministry of Environment, Science, Technology and Innovation, Republic of Ghana) (2013). *Ghana National Climate Change Policy.*

Place, F., Barrett, C.B. Freeman, H. A.,. Ramisch, J. J., and Vanlauwe, B. (2003). Pro-spects for integrated soil fertility management using organic and inorganic inputs: evidence from smallholder African agricultural systems. *Food Policy*, 28, 365–378.

Republic of Ghana (RoG) (2016). *National Climate Change Adapation Strategy.* Available at: www.adaptation-undp.org/sites/default/files/downloads/ghana_national_climate_change_adaptation_strategy_nccas.pdf. Accessed 28 March 2018.

Rosegrant, M. W., Koo, J., Cenacchi, N., Ringler, C., Robertson, R. R., Fisher, M., Cox, C. M., Garrett, K., Perez, N. D., and Sabbagh, P. (2014). *Food Security in a World of Natural Resource Scarcity: The Role of Agricultural Technologies.* Washington: International Food Policy Research Institute.

Sanchez, P. A., and Jama, B. A. (2002). Soil fertility replenishment takes off in East and Southern Africa. In: B. Vanlauwe, J. Diels, N. Sanginga and R. Merckx, eds, *Integrated Plant Nutrient Management in Sub-Saharan Africa*. Wallingford, UK: CAB International, pp. 23–45.

Vanlauwe, B., Bationo, A., Giller, K. E., Merckx, R., Mokwunye, U., Ohiokpehai, O., Pypers, P., Tabo, R., Shepherd, K. D., Smaling, E. M. A., Woomer, P. L., and Sanginga, N. (2010). Integrated soil fertility management: Operational definition and consequences for implementation and dissemination. *Outlook on Agriculture*, 39, 17–24.

Vanlauwe, B., Kihara, J., Chivenge, P., Pypers, P., Coe, R., and Six, J. (2011). Agronomic use efficiency of N fertilizer in maize-based systems in sub-Saharan Africa within the context of integrated soil fertility management. *Plant Soil*, 339: 35–50.

10 Unintended consequences of dams and water security

An insight into women's vulnerability and the spread of malaria in Ethiopia

Romy Buchner, Mridula Nair, Barbara Pinto and Larry Swatuk

Introduction

In the advent of climate change, hydrological regimes are extremely sensitive to climate variability and as the threat of anthropogenic climate change grows, this increases the variability of precipitation patterns. In response, dams have been used as an adaptation method to enhance water security and provide climate change mitigation. The construction of water-resource projects such as dams, irrigation and canals in Africa is seen as pivotal for food security and alleviating poverty, as water security may be threatened in a changing climate. However, the unintended health implications, such as the increase in vector-borne disease transmission, may minimize the intended benefits of securing water through dam construction. This chapter examines the effects of climate change, the correlation between dam construction and malaria transmission, and sheds light on the vulnerabilities of women who live in proximity to water-management projects such as Ethiopia's largest hydroelectric dam, the Gilgel Gibe III.

Climate change and its impacts on water security

Climate change and water security

Rainfall variability and its impact on water security has become a fundamental issue due to the effects of climate change on the hydrologic cycle. Globally, rising temperatures, increased water vapour in the atmosphere and changes in atmospheric circulation have contributed to increases in daily and seasonal rainfall variability (Huber and Gulledge, 2011). This variability has manifested as changes in location, type, amount, frequency, intensity and duration of rainfall, and increases in extreme events (Allen and Ingram, 2002; Trenberth 1998). Overall, these changes are causing wet regions to become wetter and dry regions to become dryer, which will have disastrous implications for water security (Skliris *et al.*, 2016; IPCC, 2007). Furthermore, in a warmer climate heavy rainfall will increase and be produced by fewer more intense precipitation events

(Trenberth, 2005). These extreme events represent an increased risk of droughts and floods.

For eastern Africa, including the Horn of Africa, temperatures will continue to rise and precipitation patterns will become more extreme if actions to slash global greenhouse gas emissions are not taken (IPCC, 2007). Originally, heavier precipitations were predicted to occur during the region's 'short rains' season from September to November, but new evidence suggests that these gains in rainfall will be offset by declining rainfall and severe dryness during the 'long rains' season from March to May. In addition, increased rainfall during the short rains is expected to result in increased surface runoff, leading to more flooding events. One such event occurred in 2006, where major floods of unprecedented spatial extent and timing occurred across Somalia, Ethiopia and other parts of East Africa (Tarhule, 2005). Evidence of these trends have been observed in several countries comprising the Horn of Africa, where overall rainfall has been continuously declining, the frequency and duration of drought have increased and droughts have been interspersed with extreme flood situations (Nicholson, 2014). The greatest increases in aridity are expected to occur in Djibouti, Ethiopia and Somalia. This will have dire implications for these regions, which are already grappling with decades of drought and extreme hunger.

Studies have found that rural areas within arid and semi-arid river basins in developing countries are most vulnerable to the effects of climate change (Millennium Ecosystem Assessment, 2005). This raises serious concerns for the 70 million people that are located in areas prone to extreme droughts within the Horn of Africa (Ndaruzaniye, 2011). Many of these inhabitants are affected directly by changes in the volume and timing of river discharge and groundwater recharge. Thus, they can be severely affected by changes in the quantity and distribution of water resources.

Increasing rainfall variability can influence anticipated variations in stream flow, lakes, reservoirs, groundwater levels and groundwater recharge. For Africa, the predicted decrease in annual precipitation over much of the continent could have severe implications for water availability. According to a study by DeWit and Stankiewicz (2006), a 10 per cent decrease in precipitation in regions that receive ~1,000 mm of rain per year would reduce surface water supplies by 17 per cent, whereas this same decrease in regions receiving ~500 mm per year would result in a 50 per cent cut. These decreases in surface water supply are projected to affect water access across 25 per cent of Africa by the end of this century. Concerns regarding water scarcity are relevant even in the instances of floods, since moderate rains soak into the soil, benefiting plants and recharging groundwater reserves, while the same rainfall amounts in a short period of time may increase the proportion that is lost through runoff, leaving soils much drier at the end of the day (Kailash *et al.*, 2014). Therefore, regions prone to flooding and/or droughts may experience water deficits if the extra water is not properly managed and stored.

In the Horn of Africa, increased scarcity and the degradation of water sources are threatening human well-being. The population in the region has increased fourfold in the past 50 years and continues to grow rapidly (UNEP, 2011). Farmers need more water to feed more people and extended areas are needed for food production. This adds pressure to the region's land and water resources. Moreover, countries within the Horn that depend strongly on hydropower for electricity generation, such as Ethiopia, will face significant challenges in maintaining their energy supply (Carius *et al.*, 2008). While Africa in general has a low adaptive capacity (Ndaruzaniye, 2011), the water security situation in the Horn of Africa is particularly severe, and has resulted in decades of periodic water scarcity and food shortages.

An ability to cope with extreme climate variability is reflected in the cultural fabric of social groups across the Horn of Africa. However, the region's governments – reflecting the artificiality of the 'modern' state form, including its built-in social inequities (Swatuk, 2012) – have been unable to develop equitable coping strategies towards the adverse impacts of increasingly extreme climate regimes. It has been estimated that over 600,000 people have died because of drought in the twentieth century alone in Ethiopia, Eritrea and Somalia (ACCES, 2010). Already today, many Eastern African countries are suffering from water stress, with large parts of the population living in poverty and not having access to clean and safe drinking water. As rainfall and surface water become less reliable, demand for groundwater has begun to grow (Calow *et al.*, 2009). However, as droughts become more intense and frequent in the region and groundwater demand increases, these sources become increasingly prone to failure (see www.faoswalim.org/ for details).

Those regions grappling with water insecurity are also faced with the challenge of growing sufficient quantities of food with limited water resources. A large part of the population within the Horn is engaged in subsistence agriculture and farm marginal lands under rainfed conditions, with relatively limited access to productive assets, inputs, technology, and services. In fact, less than 1 per cent of the cultivable area in the region is under irrigation (UNEP, 2006). Therefore, the scale and duration of rainfall is the main climatic factor determining land productivity. This dependence on rainfall has had major consequences for regional food security. During Ethiopia's 2000 and 2003 production seasons, for example, major drought affected the food security of over 10 million people, leading to episodes of famine (FAO, 2009). Climate models warn of changes that have a direct effect on grain yields and the ability of plant and animal species to survive (ACCES, 2011). These changes will have severe social and economic impacts, ultimately hindering economic stability and threatening human security.

While the combination of droughts and dependence on rain-fed subsistence agriculture has impacted the well-being and livelihoods of many Ethiopians, the recurrence of extreme weather events cannot adequately explain Ethiopia's food and water crisis. Even in years with adequate rainfall and good weather, there have always been millions of Ethiopians vulnerable to the threat of starvation. In particular, the rural poor have been disproportionately affected by food and water shortages (IFAD, 2016). In an effort to achieve its Millennium

Development Goals of attaining middle-income status by 2025, government projects and Chinese investments have stimulated economic growth that has translated in a 33 per cent reduction in the share of people living in poverty in the country (World Bank Group, 2015). Despite this achievement, at least 37 million Ethiopians continue to live in poverty and according to recent World Bank statistics, the poorest Ethiopians are becoming even poorer. This population segment is trapped at the margins of social, economic and political systems that render them chronically food and water insecure.

Ethiopian governance has historically lacked stability, which has set in motion a perverse political economy in which rules and institutions established to maximize efficiency result in unjust distributive outcomes, exacerbating inequalities (Kennedy, 2012). As Ethiopia becomes economically globalized, multinationals owning significant commercial and economic interests in Africa and domestic elites have managed to appropriate most of the surplus the economy generates. In addition, the government has consistently favoured the investment of funds into programmes that appease the urban population. For example, in 2005 a programme that allowed wheat, sugar, edible-oil and construction of low-cost condominium apartments to be subsidized prompted the government to spend hundreds of dollars on urban infrastructure and the import of luxuries, despite the fact that 81 per cent of the population live in rural areas (The World Bank, 2016). Moreover, under the current regime, land in Ethiopia is under state ownership and farmers are assigned a usufruct right to it. The government claims that the land policy is a form of social protection because it protects farmers against distress sales, land concentration in the hands of the wealthy and subsequent exploitation of the poor farmers. However, in the last few years the government has been evicting farmers to undertake urban expansion and large-scale private commercial agriculture. As the urban population and powerful corporations reap the benefits, small-scale farmers and the rural poor find themselves increasingly marginalized and struggle to cope with the rising food prices afforded by their wealthier counterparts (Swatuk, 2018).

The construction of dams as a means of adapting to and mitigating climate change have only worsened the inequalities between the urban rich and the rural poor, with the benefits of these projects being inequitably shared. The water stored within the massive reservoirs is often used as a reliable water source for large and urban communities. However, this is accompanied by water shortages for many of the adjacent rural communities (Tilt *et al.*, 2009). Often, it is the small-scale fishermen and farmers in downstream rural communities whose livelihoods are most severely impacted by river diversion and alterations in streamflow. Meanwhile, the use of dams to improve energy security has also historically benefited the urban rich. In Ethiopia, dams have been used to increase energy production, yet access to this energy is more readily available in urban areas and largely out of reach of the rural populations. Instead, most rural Ethiopians are highly reliant on biomass for cooking and heating (Virginia Tech, 2014). Therefore, by centralizing power grids, hydropower dams disproportionately benefit industries and higher income groups, widening income disparities.

The Ethiopian government's interest in economic development and the strides it has made in reducing poverty has ironically come at the expense of the most impoverished. As the rich and urban populations reap the benefits, the rural poor find themselves more vulnerable to external pressures, such as changing rainfall variability. These inequalities represent differential coping capabilities, whereby those (urban/elites) with existing economic, energy, water and food security are better equipped to respond to the impacts of climate change. Meanwhile, those (rural/peri-urban poor) with the least coping capability are rendered even more vulnerable.

Climate change and vector-borne disease

Vector-borne diseases continue to contribute significantly to the global burden of disease, and cause epidemics that have wide socioeconomic impacts, increase health inequities and act as a brake on socioeconomic development. The WHO estimates that one-sixth of the illness and disability suffered worldwide is owing to vector-borne diseases, with more than half of the world's population currently at risk. Every year, more than one billion people are infected, and more than one million people die from vector-borne diseases, including malaria, dengue, schistosomiasis, leishmaniasis, Chagas disease and African trypanosomiasis (Lozano et al., 2013; WHO, 2014). The transmission of many of these diseases is a function of the interaction with the environment, which results in varying spatial distributions and rates of transmission (Githeko et al., 2000). Given the sensitivity of these diseases to weather and climate conditions, the ongoing trends of rising temperatures and increasingly variable rainfall threaten to increase the vector-borne disease risk.

Changing temperature and precipitation may shift the geographic range in which vector-carrying mosquitoes can live and the seasonal period of disease risk. A warming climate and increases in water temperatures could allow the mosquitos to produce more offspring during the transmission period, increase transmission intensity and increase the proportion of infective vectors (Githeko et al., 2000). Increases in heavy rainfall events and associated flooding has the potential to increase the number and quality of breeding sites for vectors through its influence on standing water (Patz et al., 2003). In addition to changes in transmission rates and characteristics, higher temperatures, changes in precipitation, and climate variability will alter the geographic range and seasonality of many vector-borne diseases, which could lead to the introduction of these diseases into new non-endemic regions (Hunter, 2003). Higher temperatures in combination with increasingly variable rainfall patterns will expand the geographic range of malaria transmission to higher altitudes and latitudes and extend the transmission season in some locations (Githeko et al., 2000).

Developed and developing countries have an equal probability of being struck by the effects of climate change, though in very different ways, leading to highly varied impacts and outcomes. Clearly, limited-resource countries will be less able to cope with the associated disease outbreaks. In poor countries in particular, the impacts of major vector-borne diseases can limit or even reverse improvements in social development. The burden of climate-sensitive diseases is already greatest for these populations owing to poorer environmental and social conditions (Wu *et al.*, 2016). The effects of climate change on these diseases will only intensify this burden.

Motivations for dams

Dams have been used to manipulate water flows for thousands of years. Originally, dams were built for a single purpose, i.e. for water supply or irrigation. As civilizations developed, there was a greater need for dams' multi-purpose use: e.g. for water supply, irrigation, flood control, navigation, water quality, sediment control and energy. By 1900, several hundred large dams had been built in different parts of the world, mostly for water supply and irrigation. The century that followed saw a rapid increase in large dam building, and by the end of the twentieth century there were over 45,000 large dams in over 140 countries; representing over \$2 trillion in investments (Terrascope, 2013). According to the International Commission on Large Dams (ICOLD), an international organization that sets the standards for dams, 50 per cent of today's large dams are used for irrigation, 18 per cent for hydropower, 12 per cent for water supply, 10 per cent for flood control and the rest for other functions, including navigation, debris control and recreation (SIWI, 2012).

Irrigation and electricity are among the main purposes of dams due to their global demand and economic ties. The agricultural sector in general places the greatest demand on water resources, accounting for 70 per cent of freshwater withdrawal globally (WWAP, 2012). The vast quantities of water in reservoirs allow them to act as effective and steady sources of water for irrigation with minimal seasonal fluctuations. As of 2013, 30 per cent to 40 per cent of the 271 million hectares that are irrigated worldwide rely on irrigation dams (Terrascope, 2013). Electricity generated from dams is the largest renewable energy source in the world. World hydroelectric power plants produce over 2.3 trillion kilowatt-hours of electricity each year; generating approximately 16 per cent of the electricity consumed globally, representing 86 per cent of all electricity from renewable sources (IPCC, 2011). This energy has been promoted as 'green energy' due to its clean, renewable, non-emitting nature that provides low-cost electricity and helps reduce carbon emissions. Compared with conventional coal-power plants, hydropower prevents the emission of about $3\,\mathrm{GT}$ CO_2 per year, which represents about 9 per cent of global annual CO_2 emissions (Berga, 2016). As the world rushes into implementing the commitments enshrined in the historic climate deal in Paris, the use of large dams to mitigate climate change is becoming increasingly popular across the world.

With increasingly extreme rainfall patterns, the use of dams for water supply and flood control will continue to serve as a tool to buffer against climate change. Properly planned, designed, constructed and maintained dams contribute significantly towards fulfilling our water supply requirements. One third of the countries in water-stressed regions of the world are expected to face severe water shortages this century (World Commission on Dams, 2000). Furthermore, the uneven distribution of water supply accompanying increasingly variable rainfall patterns means that countries may have water surplus and water deficits at different temporal and spatial scales. To accommodate the variations in the hydrologic cycle, dams and reservoirs are needed to store and balance flows during different weather conditions, such as by holding flow during major flood events to prevent flooding downstream and releasing more during dry seasons to increase downstream water supply (SIWI, 2012). This ability to regulate river levels can serve as an adaptation strategy towards the increasingly frequent and severe rainfall extremes following climate change.

Although dams have many uses in the human community, they also have large impacts on the environment and populations living close to them. These impacts have both environmental and social dimensions, and encompass significant land-use changes, infrastructure development and socio-economic changes. Environmental impacts from dams can include the inundation of valuable land: habitats, productive landscapes, infrastructure, and settlements; changes to river flows, water quality, evaporation rates and sediment transport; and fragmentation of terrestrial and aquatic habitats (WWF, 2013). The primary social impacts include displacement as a result of flooding of the area used as a reservoir; loss of livelihoods, access to resources, and cultural heritage by altered river flows and ecosystem fragmentation; and threats to human health through the creation of potential breeding sites for parasites (SIWI, 2012).

A growing body of science is also increasingly suggesting that one of the dams' most attractive features – its potential to move economies towards a low-carbon future – may, in fact, be false (Fearnside, 2016). These studies have identified a number of features of dams that may ultimately be contributing to, rather than mitigating, climate change. The main concern is in regard to the organic material that flows into the reservoirs, decomposes, and emits methane and carbon dioxide into the water. Through this process, dams can end up emitting more greenhouse gases than coal-fired power plants (Washington State University, 2012). Dams also contribute to greenhouse gas emissions through the enormous amounts of energy needed to create them, as well as through the massive deforestation that precedes the flooding of the reservoir (Houghton, 2005). In addition, dams that divert water out of rivers may drain and dry up downstream wetlands that would otherwise store carbon (Kayranli et al, 2010). With this in consideration, the rapid construction of dams in the coming years will have a larger negative impact on global emissions than was previously hoped.

Their utility as both a climate change mitigation and adaptation tool has continued to be supported by the world's most elite and powerful, who are set to reap the benefits from their construction, yet challenged by the scientific community and environmental advocacy groups (see Durst *et al.* in this collection).

Despite the environmental and social costs of these engineering marvels and their associated uncertainties and contestations, their ability to propel economic growth has led, and will continue to lead, to their growing use worldwide. However, sustainable approaches to water and energy planning are required if the large-scale environmental and social impacts are to be minimized.

Dams and vector-borne diseases: overview

Dams and vector-borne diseases

Large dams are among the most potent symbols of economic development. The construction of large dams has been seen as a mechanism for promoting economic growth, ensuring food security, alleviating poverty and increasing resilience in the face of climate variability and change in sub-Saharan Africa (SSA) (Kibret *et al.*, 2015). There is an estimated 40,000 large dams, defined as impoundments more than 15 meters high or storing more than 3 million m^3 of water, and 800,000 small dams have been built, and 272 million hectares of land are currently under irrigation worldwide (Keiser, *et al.*, 2005).

As the climate changes, natural mosquito habitats in many areas of Africa are abundant even without human environmental modification. Any human alteration, such as clearing land for agriculture or construction of dams, risks exacerbating existing mosquito-associated problems by expanding, altering or creating new habitats in such a way that limited mosquito populations may explode with the availability of suitable new habitats (Norris, 2004). All mosquitoes share essentially the same pattern of biological development. Eggs are deposited on or near the surface of an existing or expected water source. Mosquitoes can develop in just a few millimetres of water, ranging from near freezing to temperatures in excess of 40°C. Habitat may include, for example, fresh or brackish water in ponds, stagnant pools, slow-moving streams, including dam reservoirs (Norris, 2004). Water-resources developments coupled with demographic changes alters human-vectorparasite–contact patterns and individuals living in proximity to infested waters have higher risk of exposure to infected mosquitoes. Mosquito-transmitted vector-borne diseases include dengue fever, rift valley fever, yellow fever, chikungunya, malaria, japanese encephalitis, lymphatic filariasis and West Nile fever (WHO, 2014, Factsheet).

As dam-building across the Global South increases in relation to hypothesized climate-induced water insecurity (Swatuk, 2018), critical questions must be asked. First, are the benefits of large-scale dams out-weighed by the costs of reservoirs as breeding grounds for vector-borne diseases? Second, are these projects undermining the overall attempt at creating water security? Third, who is being made 'water secure' and who has an increased vulnerability through these dam-building exercises? The unintended negative consequences in health costs related to dams built in a precarious climate may indicate that dams are not the method in which water security can be achieved. More than one million people die every year from the direct causes of malaria. The additional costs can be measured in an estimated loss of 46.5 million disability-adjusted life-years (DALYs) with almost 90 per cent currently concentrated in sub-Saharan Africa (Keiser *et al.*, 2005). As dams and irrigation schemes transform ecosystems, they substantially change the nature of malaria risk in proximity to their location. These factors are particularly important in the establishment and operation of resource development projects. In Africa alone, 9.4 million people live near to large dams. Studies have indicated the correlation of proximity to dams between three to five kilometres as an indicator of higher risk factors for transmission of malaria (Kibret *et al.*, 2015).

There is strong correlation between large dams and the transmission of vector-borne illness, particularly malaria. In a research review conducted by Kibret (Kibret *et al.*, 2015), in areas of unstable transmission within sub-Saharan Africa, approximately 919,000 malaria cases per year were associated with the presence of 416 dams. The concluded research included a total of 15 studies on 11 dams, and investigated the impact of dams on malaria in semi-arid areas and highland fringes with seasonal malaria transmission in Africa. These studies outline that malaria prevalence was higher in dam villages than non-dam villages. In areas of stable malaria transmission, 204,000 malaria cases per year were associated with the presence of 307 dams. The data also suggest that malaria cases in areas of unstable malaria transmission were on average 3.2 times greater in communities living close to existing reservoirs than those living more than 5 km from them. Overall, the reservoirs investigated account for 0.6 per cent of the total malaria burden in the SSA. However, in the vicinity of the reservoirs in stable and unstable areas, reservoirs associated with large dams contribute to 47 per cent of malaria cases on average in communities living within 5 km (Kibret *et al.*, 2015).

Ethiopia's hydropower: the Gilgel Gibe Dam III

Ethiopia: hydro-dams politicized

As previously discussed, dams have negative ecological and social implications. The promise of economic growth through export potential is a prominent justification for dam-building in Ethiopia, which has one of the world's lowest rates of access to modern energy services (Hathaway, 2008). Ethiopia is one of the enthusiastic participants in massive dam building, with the Grand Ethiopian Renaissance Dam (GERD) being the primary contemporary example (Cascão and Nicol, 2016; see also Water International Vol. 41 No. 4 entitled 'The Grand Ethiopian Renaissance Dam: legal, political and scientific challenges'). There are three overarching goals for Ethiopia's dam projects: to boost domestic electricity production and consumption; to reduce Ethiopia's vulnerability to climate and hydrology, weakening its dependence on erratic rain-fed cultivation, enabling irrigated production for both internal consumption and international exports; and, to benefit economically by exporting hydro to neighboring countries (Verhoeven, 2013). In 2005, the Ethiopian Power System Expansion Master Plan (EPSEMP) 2006–2030 was motivated to increase hydro domestic demand in five years to nine countries to become potential buyers of electricity generated in Ethiopia: Djibouti, Egypt, Eritrea, Kenya, Somalia, Somaliland, South Sudan, Sudan and Yemen (Cuesta-Fernández, 2015).

Eight hydropower dams account for over 85 per cent of Ethiopia's existing 767 MW generating capacity and five additional hydropower sites with a combined capacity of 3,125 MW are currently under construction (Yewhalaw *et al.*, 2009). Construction costs of GERD are estimated at US$4.8 billion, funded from local taxes, donations and government bonds (UN Africa Renewal, 2014). Ethiopians abroad and at home contributed the first $350 million, with government workers contributing amounts equivalent to a month of their salaries (UN Africa Renewal, 2014).

Ethiopia's current generation facilities, including the Grand Renaissance dam and the series of Gilgel Gibe dams, have increased Ethiopia's production capacity to a level more than three times above that of the country's current domestic demand. So, while dams are justified as a development tool to alleviate poverty, the fact that only 17 per cent of the population has access to electricity presents Ethiopia's government with a 'grand challenge' to use revenue generated by the sale of hydropower beyond the country's borders to extend access within its borders and across the entire population (see Hathaway, 2008).

Dams as pathways for development is a highly contested topic, as dams for hydropower is highly politicized in Ethiopia for a number of reasons. Dam creation is often tied with land access and large-scale foreign land acquisitions, or 'land grabbing' (Swatuk, 2018). Populations living on the banks of the Omo River downstream from the Gibe III reservoir will no longer benefit from flood retreat cultivation, which is essential to their livelihoods. At the same time, this stored water is intended for use in large-scale irrigation on lands to be given out to foreign investors (Abbink, 2012). This is a clear example of 'land grabbing', or more accurately, given water's intrinsic relationship to both land (for food and other crop production) and energy, 'water grabbing', on the site of the Gilgel Gibe III (Swatuk and Cash, 2018; Franco *et al.*, 2013).

The desire for rapid construction of hydro-dams is argued to be deeply rooted in the modernized development ideology, as well as a mechanism to increase state-building. Through the power and prestige of stimulating private sector and smallholder initiatives, the political inner circle is increasingly positioned to firmly control the development process (Verhoeven, 2013). This benefits select key players and boosts national prestige simultaneously, as it deepens the model of formal and informal power at the state level. The Ethiopian government, led by the Ethiopian People's Revolutionary Democratic Front (EPRDF), views dams as the most proactive way to attain energy and food self-sufficiency as well as to ensure water security thereby empowering a sovereign Ethiopia (Abbink, 2012): hence the notion of 'renaissance' in the GERD project. Verhoeven (2013) describes this dam-building renaissance as tantamount to a hydro-agricultural state-building project, so, too, highlighting how dynamic and contradictory Africa's changing politics of water security and energy development are at the national, local and regional levels (Swatuk, 2012).

Case study: Gilgel-Gibe hydroelectric dam

As part of the East African Power Pool (EAPP) initiative that was launched in 2005 to facilitate the trade of electricity between countries (International Rivers, 2011), Ethiopia built the Gibe III hydroelectric dam on the Omo River, which supplies more than 80 per cent of the inflows to Lake Turkana (Velpuri and Senay, 2012). Investment in the project was estimated to be about US$2.11 billion, of which the Ethiopian government provided US$572 million from its national budget (International Rivers, 2011). The remaining funds were secured from Industrial and Commercial Bank of China (ICBC), which supports the Ethiopian Electric Power Corporation through a US$500 million loan. This loan was essential for Gibe III after the World Bank, African Development Bank and European Investment Bank withdrew, due to social and environmental concerns (Verhoeven, 2013).

Evidence shows that the highest environmental and social costs stemming from the construction and operation of the Gilgel Gibe III project are paid by the most vulnerable. To aid in the assessment of potential impacts, as required by Ethiopian and international law, the building of dams requires environmental and social impact assessments (ESIA) to evaluate the potential harm of implementing the infrastructure. In the case of the Gilgel Gibe III, the assessments were not conducted in a responsible time-frame to fully evaluate the potential challenges to local communities and the natural environment. Controversially, the results of the assessment were released two years after the construction of the dam had started and with all plans already determined and approved by the state authorities (Abbink, 2012).

According to International Rivers (2011), by eliminating the Omo River's natural flood cycle, the Gilgel Gibe III brings major hydrological changes to a very fragile ecosystem, and puts the Dassanech, Mursi, Nyangatom and other indigenous peoples at great risk. At least 100,000 people depend on food cultivated in the river's flooded banks, a practice known as flood-retreat cultivation and the river's harvest helps support an additional 100,000 people through local trading practices between farmers and herders (International Rivers, 2011). This traditional food system is crucial for these communities because they live in one of the poorest and remotest parts of Ethiopia and have long been politically marginalized (International Rivers, 2011).

Unintended consequences

The problem of malaria is very severe in Ethiopia, where it has been the major cause of illness and death for many years. According to records from the Ethiopian Federal Ministry of Health, 75 per cent of the country is malarious, with about 68 per cent of the total population, or over 54 million people, living in areas at risk of malaria (Campbell-Lendrum *et al.*, 2017; Adhanom *et al.*, 2006). This population experiences major epidemics every five to eight years, but focal epidemics occur every year. The transmission risk of malaria is dependent on a range of environmental and socioeconomic factors. In Ethiopia, malarial transmission is unstable, seasonal and depends primarily on altitude and rainfall. Exposure typically peaks from September to December and April to May, which coincide with the harvesting season for farmers (Alemu *et al.*, 2012). Due to the unstable and seasonal pattern of transmission, the protective immunity of the population is generally low, and all age groups are at risk of infection and disease (Ayele *et al.*, 2012). However, individuals with poor socio-economic conditions, particularly children and women, are at greater risk of infection (Legesse *et al.*, 2007).

The region surrounding the Gilgel Gibe III Dam encompasses a range of altitudes, and is subsequently subject to distinct malarial distributions. The highest peaks are affected by occasional transmission, while the highland fringes range from high to low transmission and the lowlands with intense transmission (Alelign and Dejeune, 2016). Regional studies have shown that at the Gilgel Gibe III site, rainfall and relative humidity are driving forces behind malaria (Sena *et al.*, 2015). This is of concern given the anticipated changes in these variables as a result of climate change. As with the rest of Eastern Africa, temperatures are expected to warm in Ethiopia, increasing by up to 5.1°C by the 2090s (McSweeney *et al.*, 2010). Rainfall, however, has been more difficult to project. Attempts to model future changes in rainfall show large ranges relative to the baseline, alluding to possible increases and decreases in annual amounts (Ababa, 2007). However, it is generally expected that there will be an increase in the intensity of extreme precipitation events, which will increase flood risk and mosquito breeding sites.

Through the changes in temperature and rainfall patterns and the subsequent effects on the distribution and seasonal transmission of malaria, an overall increase in malarial transmission is anticipated across the Ethiopian landscape. The disease is projected to show increases of more than 100 per cent in person-months of exposure in Ethiopia towards the end of the twenty-first century (Tanser *et al.*, 2003). In addition, parts of the highlands are expected to become newly suitable for malaria by 2050 (Rogers and Randolph, 2000). As a result, more than 122 million Ethiopians will be exposed to the disease (Tanser *et al.*, 2003). These projections compound the concerns regarding the regional effect of the Gilgel Gibe III Dam on malarial transmission and must be considered when planning for the potential impacts of future water-resource development projects.

Due to recent Gilgel Gibe III Dam construction, large numbers of people living in villages in proximity are at potentially greater risk of malaria transmission in this area. For example, a high number of Anopheline arabiensis (the major malaria vector) was recorded in the Gilgel Gibe area (Yewhalaw *et al.*, 2014). Studies conducted on malaria transmission patterns reveal that increased risks of exposure to malaria is correlated with village proximity (within 3 to 5 kms) to the dam reservoir. A two-year study indicated that among children under 10 years of age living in 16 study villages around the Gilgel Gibe, exposure to malaria varied directly with proximity to the dam. This study indicated that mosquito density decreased by 53 per cent at 6 to 7 km from the dam, compared with localities close to the dam reservoir (Yewhalaw *et al.*, 2013). The findings also indicated that malaria prevalence was 7.7 per cent in communities within 3 km as compared to 4.4 per cent in those living more than 3 km from the dam (Yewhalaw *et al.*, 2013).

Women's vulnerabilities

Women's vulnerability to vector-borne diseases may be evident; however, limited information on why women and their children have increased rates of malaria is often less discussed. It is argued here that there are biological, social and environmental determinants that increase women's risk for transmission of vector-borne diseases. Since women carry the high burden of vector-borne infections, it is important to investigate why women are more vulnerable than men.

A gender analysis is encouraged when analyzing the epidemiology of vector-borne diseases, as the understanding of transmission is governed by epidemiology; however, prevention and treatment are affected by gender stratification in societies. The biological determinants are specific to pregnant women, making them more at risk. Studies have indicated that pregnant women have lowered immunity to malaria, which leads to anemia, miscarriages, low-birth rates, neonatal deaths and maternal deaths. Extreme poverty only heightens the vulnerability for this segment of the population. Approximately 125 million pregnant women are infected by malaria each year. Malaria is known to cause 75,000–200,000 infant deaths yearly in sub-Saharan Africa, and a large proportion of maternal deaths (37%) during pregnancy are attributable to malaria in several African countries. In endemic areas, malaria can cause placental malaria and severe, malaria anemia. Maternal anemia infection might lead to intra-uterine growth restriction and prematurity among newborn, which in turn may lead to low birth weight (Austin *et al.*, 2014).

Pregnant women living near the Gilgel Gibe III are particularly at risk. A high prevalence of anaemia indicates it is currently a serious health problem of pregnant women living in Gilgel Gibe Dam area. In a study conducted to assess risk factors of living in proximity to the dam reservoir, anemia was found in 53.9 per cent of the 388 pregnant women tested (Getachew, *et al.*, 2012). It has been suspected that nutrition might influence susceptibility to infection by the malaria parasite or modify the course of disease. There have been relatively few efforts to examine such interactions. In most countries in sub-Saharan Africa, more than 20 per cent of the women are classified as being malnourished. Malnourished pregnant women are more likely to give birth to underweight babies who then in turn are more susceptible to infectious disease including malaria (Rylander *et al.*, 2013).

Social indicators, which include economic and cultural circumstances in which men and women live, contribute to women's vulnerabilities by influencing their access to preventative measures and treatment services.

A gender-perspective framework is important in recognizing health-seeking behaviours, both at prevention and treatment levels. Early recognition of symptoms can be the key to addressing women's vulnerability to adverse outcomes. When the gender-analysis literature is examined, the common theme expressed in understanding women's vulnerabilities is the overarching socio-cultural indicator of women's lack of autonomy. This has some of the largest consequences on women's health status. Gender inequalities, including economic, health and education, can have an impact on women's malaria rates. Women's socio-health status has the strongest direct correlation with malaria rates and overall health outcomes measured by women's legal economic status, formal access to land, access to loans and access to property for women, with increased social status being correlated with lower rates of malaria (Austin *et al.*, 2014).

In rural areas, women's gendered domestic roles increase their vulnerabilities, as women carry the major responsibility for the well-being of the household, although this responsibility is rarely matched by autonomy to make decisions or by access to the necessary resources (Tanner and Vlassoff, 1998). In both health prevention and treatment-seeking behaviours, women may have to ask permission from their husbands, mothers-in-law or senior household males before being permitted to seek care. In some cultures, women cannot visit health centres unaccompanied. This may add additional barriers for women to act upon their desire to seek treatment. Women's lack of financial autonomy suggests that women have an inability to make decisions about how money is spent in the household; therefore, they are limited in the ability to prioritize preventive measures, such as ensuring use of bed nets within their household. Revenues earned by women are more often used to meet basic needs that improve quality of life, such as education fees, health care costs, clean water and sanitation services, and clothing for children, in comparison to earnings made by men (Austin *et al.*, 2014). Although women carry the largest burden of malaria, whether confronted with a child with the disease or herself, women in most societies tend to be the ones that are often poorly informed about disease risks and possibilities for prevention and cure (Tanner and Vlassoff, 1998). The lack of knowledge of exposure, symptoms and treatment may correlate with women's inequalities in education. In rural Ethiopia the literacy rate for rural women is just 19 per cent, compared with 43 per cent for men (Jones *et al.*, 2010).

Domestic and reproductive gender roles such as child care, domestic chores and subsistence agriculture link women to water resources for domestic needs. Since women and girls often cook, clean, farm and provide health care and hygiene for their households, they are on the front line of their communities' and countries' water issues. Women also play a crucial role in ensuring the availability of water for the household's survival (Fonjong and Fokum, 2015). Therefore, environmental challenges due to climate change also increase women's vulnerabilities as they are directly related to water scarcity. With livelihoods being directly related to natural resources, women are particularly at risk from the predicted impacts of climate change on water resources (Baker *et al.*, 2015).

Women's domestic water use may also be challenged by global water-related issues such as over-consumption, population growth, water privatization and climate change, all of which affect the quality and accessibility of water. As women are primarily responsible for securing water for domestic use, they will be forced to adjust to such changes (Fonjong and Fokum, 2015). In addition, water scarcity and contamination disproportionately impact low-income women, and their health-seeking opportunities are further jeopardized for many who must walk miles to access clean water. Clean-water shortages and man-made infrastructures such as dams may require women to walk further distances in order to access water resources. Access to water, sanitation and hygiene are indicators of good health outcomes, as it is primarily a woman's domestic chore to gather water. The ability to adequately implement these practices may be threatened not only by inadequate accessible sources of clean water due to rain-fall variability, but also the risk of exposure to high densities of vector-borne carriers in her attempt to secure water as she is forced to walk near mosquito habitats.

Effective interventions for water security and malaria

The way forward

Reduced access to natural resources for remote communities after dam construction such as land, water and forests often negatively impacts their livelihoods, since natural resources represent, for the majority of those people, the main capital asset (Velpuri and Senay, 2012). Not only are large dams costly and prone to systematic and severe budget overruns, they also take a long time, on average 8.6 years, to be built (Ansar et al., 2014). Climate change is believed to further exacerbate existing vulnerability to disease and food insecurity for remote populations since they are more reliant on agriculture, are more vulnerable to droughts and have a lower adaptive capacity relative to their urban-rich counterparts (Huynen et al., 2013).

The key to successful governance and overcoming the threats to water security is institutional change that brings together key stakeholders in ways that ensure long-term, sustainable futures for the environment, water users and their communities (Connell and Grafton, 2011). This section will shed light on the preventative measures that can be implemented to combat malaria and ensure water security in Ethiopia.

Potential solutions

As discussed, the use of dams for climate change mitigation will likely continue to grow despite the negative outcomes, as witnessed in Ethiopia. Alternative solutions are difficult to integrate into the dam-for-development debate, which rationalizes the negative social and ecological outcomes under the veil of economic growth. There are few realistic or conceivable approaches to address both the adaptation and mitigation strategies of climate change as they relate to water security. The following solutions may not be effective in Ethiopia's context, as the government is determined to bring growth and 'power' to Ethiopia through the state-building multi-dam schemes. However, they may shed light on alternative ways to move forward.

Alternative energy

Ethiopia's dependence on hydropower creates potential risks, as their water security method is highly vulnerable to droughts. First, alternative green energy sources need to be further investigated for the country's domestic energy demand. Ethiopia needs to diversify its energy generation methods. Solar power has significant potential in Ethiopia. There may be great potential for Ethiopia to convert their dependency on hydropower to innovative solar water-heating and large-scale grid plants. There is also the possibility of at least 30 MW of power from cogeneration in some of the country's sugar factories (Hathaway, 2008). In delivering 'electricity to the poor', the government would be wise to invest revenue generated by the cross-border sale of hydropower into small-scale renewables for the rural masses.

Rainwater harvesting

Rainwater harvesting creates synergies by upgrading rain-fed agriculture and enhancing productive landscapes. The rainwater use by crops and natural vegetation is in many cases by-passed in integrated water resource management (UNEP, 2009). Consequently, the rainwater harvesting interventions are not widely recognized in water policy or in investment plans, despite the broad base of cases identifying multiple benefits for development and sustainability (UNEP, 2009). By introducing policies recognizing the value of ecosystem services and the role of rainfall to support these systems, rainwater harvesting emerges as a set of interventions addressing multiple issues on human well-being and improved ecosystems services. Governments and communities should jointly make efforts in enabling policies and legislation, together with cost sharing and subsidies for rainwater harvesting interventions.

To be sure, dams themselves are a means of 'rainwater harvesting', but in our view, rural food and livelihood security is more likely to be achieved through the rollout of rainwater harvesting systems for domestic use as well as small, 'one farm–one dam' or 'one community–one dam' water-storage systems.

Linking multiple dam operations

While making less of a political statement – where large dams are an obvious expression of state power – multiple smaller dams arranged in a cascade can transfer water between reservoirs as needed and be far more effective than large dams in delivering desired social, economic and environmental outcomes. Such an interconnected system would minimize evaporation losses by storing water in the most upstream reservoir and transferring water to reservoirs downstream for water supply only when needed, thus minimizing the need for unnecessary water storage and subsequent water losses (Watts *et al.*, 2011). Aside from operating as large evaporation pans, dams in low-lying areas are a breeding ground of disease vectors. Those located upstream in cooler environments are less likely to support disease vectors such as mosquitos.

Mitigation measures

Aside from alternative green energy sources, potential solutions must also focus on alternative mitigation measures to address the adverse effects of large-scale dams. There are many options for mitigating the negative environmental and social impacts of dams, although they are rarely implemented primarily due to financial restrictions. As discussed, one of the major concerns with large dams is the potential downriver hydrological changes. These impacts can be minimized, for example, by managing the water releases from the turbines or 'reoperation'. Hydro-dams, which generate baseload electricity, such as the Gilgel Gibe III, are potentially capable of mimicking downriver flows to provide adequate down-river water supply through off-channel pumped storage. Alternatively, coord-inated operations within a cascade of dams can address extreme daily fluctuations in flow and assist in restoring the natural seasonal patterns (Watts *et al.*, 2011). This water-flow management can help offset the impacts of the dam on riparian ecosystems, reservoir and downriver water quality, and aquatic weed and disease vector control (Ledec and Quintero, 2003).

It can also ensure enough water for small farmers and communities downstream.

Highly eutrophic dam reservoirs are potentially challenged by floating aquatic vegetation, which can cause problems such as a loss of habitat for fish and other aquatic species, which may lead to extinction. Floating vegetation has also been linked to the creation of breeding grounds for mosquitos and other disease vectors. Therefore, pollution control and pre-impoundment selective forest clear-ing and the occasional drawdown of reservoir water-levels, may be used to kill aquatic weeds (Ledec and Quintero, 2003). As an innovative way to reduce the eutrophication in dam reservoirs, an Artificial Floating Island (AFI) can be added to the reservoir. This mechanism uses a small buoyant frame on which plants grow; they act to compete with algae for excess nutrients in the water. They can also reduce biochemical and chemical oxygen demands by the water up to 60 per cent (Kamble and Patil, 2012), although further research on this method is needed to support its potential to mitigate the adverse effects on the ecosystem.

Conclusive and timely Environmental and Social Impact Assessments (ESIA) must be conducted for each new proposed large-scale dam or irrigation scheme. Unfortunately, with the siting of the Gilgel Gibe III, this was not the case. Each proposed dam should also include an implementation of mitigation measures for cumulative impacts of multi-dam schemes. These should be completed prior to the construction of any additional dams. Increased research on gender analysis as it relates to climate change is needed to fully understand how this increases women's risks for life-threatening diseases. Women's domestic relationship with water management may also increase their risk to exposure. As a consequence of climate change, effects such as the depletion of natural resources will lead men and women to interact with natural resources and landscapes in different ways. It is imperative that the ESIA include women's specific vulnerabilities as criteria in order to avoid potential negative outcomes.

Management for malaria/vector borne diseases

In the era of sustainable development across diverse sectors, water, sanitation and hygiene (WASH) forms the core and needs to be prioritized. Despite the progress seen in Ethiopia, 43 per cent of the population does not have access to an improved water source and 28 per cent practice open defecation while the majority of health facilities lack access to clean water and only about 32 per cent have access to safe water (WHO, 2014). The disease burden related to insufficient and inadequate WASH facilities is evident, with deaths every year attributed specifically to these health determinants (Bartram, Lewis, Lenton and Wright, 2005). The economic liabilities associated with mortality and malnutrition due to unsafe WASH, time lost in collecting water, and seeking somewhere to defecate is of major concern (Bartram and Cairncross, 2010). Combating health related issues in accordance with WASH in rural communities calls for local solutions that includes participatory approaches. Community level knowledge and practices related to the ecology and environmental management for disease control could result in lasting sustainable reductions of vector-borne diseases such as malaria (Randell et al., 2010).

Community health workers (CHWs) as an intervention in the health and WASH sectors play a crucial role in broadening access and coverage of health services in remote areas and can undertake actions that lead to improved health outcomes (WHO, 2014). Environmental management practices for disease control can be implemented at the community level by mobilizing local resources through financial aid and educational aid to community health workers (Randell et al., 2010). Governments, NGO's and foundations need to strengthen their engagement to accelerate the provision of WaSH amenities to rural remote communities. Inclusion and participation of women in stakeholder decision-making processes at all levels should be one of the top key objectives of global health national and international leaders and practitioners. CHWs with adequate training and support offer simple and effective ways to increase uptake of malaria prevention and effectively deliver essential healthcare (IPTp and ITN services) to women of reproductive age, as well as counseling to communities (Okeibunor et al., 2011). CHWs play an important role in the redistribution of health service provision to less specialized health workers in order to bring essential health services closer to populations with limited or no access to essential public health services (Buchner *et al.*, 2014).

There is good evidence that insecticide treated nets (ITNs) are a cost-effective and successful method of reducing malaria morbidity and mortality and their use provides a protective efficacy against malaria of up to 60 per cent, and in areas with stable malaria they reduce the incidence of uncomplicated malarial episodes by 50 per cent (Williams *et al.*, 2009). Furthermore, ITNs can avert more than just malarial deaths—in children, their use has been shown to significantly reduce under-five childhood mortality due to all causes (Williams *et al.*, 2009). Mass, free ITN distribution campaigns and subsidized voucher programmes for ITNs targeting pregnant women and children under five years of age have been successful at increasing the proportion of households that own and use an ITN (Sangaré *et al.*, 2012).

Despite the considerable increase in funds over recent years to control malaria in Ethiopia, the disease has been the most frequently reported cause of morbidity and mortality as intermittent preventive treatment (IPT) for malaria prevention during pregnancy has not been adopted completely (Ali and Deressa, 2009). The availability and affordability of IPT for community-based malaria control interventions carried out by village-based community health workers (CHWs) and home-based management of malaria is a major concern (Ali and Deressa, 2009). IPT prevents the adverse consequences of malaria on maternal and fetal outcomes, such as placental infection, clinical malaria, maternal anaemia, fetal anaemia, low birth weight and neonatal mortality and has been shown to be highly cost-effective for both prevention of maternal malaria and reduction of neonatal mortality in areas with moderate or high malaria transmission (WHO, 2014).

Scaling up the interventions of sustainable distribution of ITNs and educating women in rural communities through outreach programmes on the benefits of ITN's and IPT is crucial for the prevention and cure of the disease. Malaria-control programmes should strive to achieve full protection in pregnant women by both IPTs and ITNs to maximize their benefits.

Conclusion

Hydro-dams continue to be regarded as important mechanisms to enhance state-building through economic growth, agricultural development and foreign trade. Yet, as highlighted in this chapter, throughout the high-modern dam-building era there are clear winners and losers. Gilgel Gibe III will no doubt pave the way for particular forms of 'development' in Ethiopia, but it also places many people at risk. 'Land grabbing' and 'water grabbing' by foreign entities indicate that Ethiopia's impoverished populations, particularly women as the most vulnerable affected by the building of the Gilgel Gibe III, are not a priority in the 'development' plan of Ethiopia. The loss of capacity to ensure healthy, safe and thriving livelihoods has been threatened as Ethiopia exploits hydro potential. As shown above, climate change will be a major contributor to the spread of malaria across sub-Saharan Africa. At the same time, dams envisioned as both climate change mitigation and adaptation mechanisms threaten the health and livelihoods of millions of people, in particular poor, rural women. Given the network of powerful local, regional and global actors aligned to 'nexus' thinking regarding state security in relation to water, energy and food, it is unlikely that dams will be reconsidered as viable development options any time soon. What that leaves us with, then, are a wide array of small-scale, local-level, 'triage'-oriented actions in the service of rural health and water security: WASH programmes, treated bed nets, household scale rainwater harvesting, perhaps a few community-level water-storage supply systems.

References

Ababa, A. (2007). Climate change national adaptation programme of action (Napa) of Ethiopia. National Meteorological Services Agency, Ministry of Water Resources, Federal Democratic Republic of Ethiopia.

Abbink, J. (2012). Dam controversies: contested governance and developmenta discourse on the Ethiopian Omo River dam. *Social Anthropology*, 20(2), 125–144.

Adhanom, T. D. W., Witten, H. K., Getachew, A., Seboxa, T. (2006). *The Epidemiology and Ecology of Health and Disease in Ethiopia* (1st edition). Addis Ababa, Ethiopia: Shama PLC, pp. 556–576.

Africa, Climate Change, Environment and Security (ACCES) (2010). Climate change and security in Africa: Vulnerability Report. Retrieved on 31 March 2017 from www.gwiwater.org/sites/default/files/pub/ACCES%20Vulnerability%20Report_1.pdf

Africa, Climate Change, Environment and Security (ACCES) (2011). Climate change and security in Africa: Vulnerability discussion paper. Retrieved on 31 March 2017 from www.africa-eu-partnership.org/sites/default/files/documents/doc_climate_vulnerability_discussion_paper.pdf

Alelign, A., and Dejene, T. (2016). Current status of malaria in Ethiopia: Evaluation of the burden, factors for transmission and prevention methods. *Acta Parasitologica Globalis*, 7(1), 01–06.

Alemu, A., Muluye, D., Mihret, M., Adugna, M., and Gebeyaw, M. (2012). Ten year trend analysis of malaria prevalence in Kola Diba, North Gondar, Northwest Ethiopia. *Parasites and Vectors*, 5, 173.

Ali, A., and Deressa, W. (2009). Malaria-related perceptions and practices of women with children under the age of five years in rural Ethiopia. *BMC Public Health*, 9, 259–259.

Allen, M. R., and Ingram, W. J. (2002). Constraints on future changes in climate and the hydrologic cycle. *Nature*, 419, 224–232.

Ansar, A. Flyvbjerg, B., Budzier, A. and Lunn, D.(2014). Should we build more large dams? The actual costs of hydropower mega project development. *Energy Policy*, 69, 43–56.

Austin, K. F., Noble, M. D., and Mejia, M. T. (2014). Gendered vulnerabilities to a neglected disease: A comparative investigation of the effect of women's legal economic rights and social status on malaria rates. *International Journal of Comparative Sociology*, 55(3), 204–228.

Ayele, D. G., Zewotir, T. T., and Mwambi. H. G (2012). Prevalence and risk factors of malaria in Ethiopia. *Malaria Journal*, 11(1), 195.

Baker, Tracy J., Cullen, B., and Abebe, Y. (2015). A socio-hydrological approach for incorporating gender into biophysical models and implications for water resources research. *Applied Geography*, 62, 325–338.

Bartram, J., Lewis, K., Lenton, R., and Wright, A. (2005). Focusing on improved water and sanitation for health. *The Lancet*, 365(9461), 810–812.

Bartram, J., and Cairncross, S. (2010). Hygiene, sanitation, and water: Forgotten foundations of health. *PLoS Medicine*, 7(11), e1000367.

Berga, L. (2016). The role of hydropower in climate change mitigation and adaptation: A review. *Engineering*, 2(3), 313–318.

Buchner, D. L. Brenner, J. L., Kabakyenga, J., Teddy, K., Maling, S., Barigye, C., Nettel-Aguirre, A., and Singhal, N. (2014). Stakeholders' perceptions of integrated community case management by community health workers: A post-intervention qualitative study. *PLoS One*, 9(6).

Calow, R. C., MacDonald, A. M., Nicol, A. L., and Robins, N. S. (2009) Groundwater security and drought in Africa: linking availability access and demand. *Ground Water*, 48(2), 246–256.

Campbell-Lendrum, D., Manga, L., Bagayoko, M., and Sommerfeld, J. (2017). Climate change and vector-borne diseases: what are the implications for public health research and policy? *Philosophical Transactions Research Society*, 370, 20130552.

Carius, A., Tänzler, D., and A. Maas. (2008). *Climate Change and Security: Challenges for German Development Cooperation.* Eschborn: GTZ. Retrieved on 31 March from www.preventionweb.net/files/8023_enclimatesecurity1.pdf

Cascão, A. E., and Nicol, A. (2016). GERD: new norms of cooperation in the Nile Basin? *Water International*, 41(4), 550–573.

Connell, D. and Grafton, Q. (2011). Water reform in the Murray-Darling Basin. *Water Resources Research*, 47(12).

Cuesta-Fernández, I. (2016). Mammoth dams, lean neighbours: Assessing the bid to turn Ethiopia into East Africa's powerhouse. In: *A New Scramble for Africa?: The Rush for Energy Resources in Sub-Saharan Africa.* London: Routledge, pp. 93–110.

De Wit, M., and Stankiewicz, J. (2006). Changes in surface water supply across Africa with Predicted climate change. *Science*, 311(5769), 1917–1921.

Fearnside, P. M. (2016). Brazil's Amazonian forest carbon: the key to Southern Amazonia's signifcance for global climate. *Regional Environmental Change* (July): DOI 10.1007/s10113-016-1007-2.

Food and Agriculture Organization of the United Nations (FAO) (2009). The state of food insecurity in the world 2009: Economic crises – impacts and lessons learned, (Rome, 2009). Retrieved on 31 March 2017 from ftp://ftp.fao.org/docrep/fao/012/i0876e/i0876e.pdf

Franco, J., Mehta, L., and Veldwisch, G. J. (2013). The global politics of water grabbing. *Third World Quarterly*, 34(9), 1651–1675. http://ftp.fao.org/docrep/fao/012/i0876e/i0876e.pdf

Fonjong, L. N., and Fokum, V. Y. (2015). Rethinking the water dimension of large scale land acquisitions in sub-Saharan Africa. *Journal of African Studies and Development*, 7(4), 112–120.

Getachew, M., Yewhalaw, D., Tafess, K., Getachew, Y., and Zuynudin, A. (2012). Anaemia and associated risk factors among pregnant women in Gilgel Gibe dam area, Southwest Ethiopia. *Parasites and Vectors*, 5(1), 296.

Githeko, A. K., Lindsay, S. W., Confalonieri, U. A., and Patz,, J. A. (2000). Climate change and vector- borne diseases: a regional analysis. *Bulletin of the World Health Organization*, 78(9).

Hathaway, T. (2008). What cost Ethiopia's dam boom?: A look inside the expansion of Ethiopia's energy sector. *International Rivers: People, Water, Life*. Retrieved from www. internationalriversorg/files: r EthioReport06Feb08.   pdf.

Houghton, R. A. (2005). Tropical deforestation as a source of greenhouse gas emissions. *Tropical Deforestation and Climate Change*, 13.

Huber, D., and Gulledge, J. (2011). Extreme weather and climate change. Centre for climate and energy solutions. Retrieved on 31 March 2017 from www.c2es.org/publications/extreme-weather-and-climate-change

Hunter, P. R. (2003). Climate change and waterborne and vector-borne disease. *Journal of Applied Microbiology*, 94(s1), 37–46.

Huynen, M. M. T. E., Martens, P., and Akin, S.-M. (2013). Climate change: an amplifier of existing health risks in developing countries. *Environment, Development and Sustainability*, 1–18.

Intergovernmental Panel on Climate Change (IPCC) (2007). Fourth assessment report: The physical science basis. Chapter 11, p. 868. Retrieved on 23 March from www.ipcc.ch/pdf/assessmentreport/ar4/wg1/ar4-wg1-chapter11.pdf

International Fund for Agricultural Development (IFAD) (2016). Investing in rural people in Ethiopia. Retrieved on 17 April 2017 from www.ifad.org/documents/10180/d030748c-44d6-455e-a35b-5f35ac8893e0

International Rivers (2011). Ethiopia's Gibe III Dam: Sowing hunger and conflict. Available at: www.internationalrivers.org/resources/ethiopia-s-gibe-iii-dam-sowing-hunger-and-conflict-2643 accessed 16 May 2018.

Jones, N., Tafere, Y., and Woldehanna, T. (2010). Gendered risks, poverty and vulnerability in Ethiopia: To what extent is the Productive Safety Net Programme (PSNP) making a difference. London: Overseas Development Institute. www. odi. org/sites/odi. org. uk/files/odiassets/publications-opinion-files/6250. pdf .

Kailash, R., O'Leary, T., Turner, T., Petrakis, G., Leonard, M., and Westra, S. (2014). Changes to the temporal distribution of daily precipitation. *Geophysical Research Letters*, 41:8887–8894.

Kamble, R., and Patil, D. (2012). Artificial floating island: solution to river water pollution in India. Case study: rivers in Pune City. *International Conference on Environmental, Biomedical and Biotechnology IPCBEE* (Vol. 41).

Kayranli, B., Scholz, M., Mustafa, A., and Hedmark, Å. (2010). Carbon storage and fluxes within freshwater wetlands: a critical review. *Wetlands*, 30(1), 111–124.

Keiser, Jennifer, De Castro, M. C., Maltese, M. F., Bos, R., Tanner, M., Singer, B. H., and Utzinger J. (2005). Effect of irrigation and large dams on the burden of malaria on a global and regional scale. *The American Journal of Tropical Medicine and Hygiene*, 72(4), 392–406.

Kennedy, D. (2012). African poverty. *Washington Law Review*, 87, 205.

Kibret, Solomon, Lautze, J., McCartney, M., Wilson, G. G., and Nhamo, L. (2015). Malaria impact of large dams in sub-Saharan Africa: maps, estimates and predictions. *Malaria Journal*, 14(1), 339.

Ledec, G., and Quintero, J. D. (2003). Good dams and bad dams: Environmental criteria for site selection of hydroelectric projects. The World Bank. Latin America and Caribbean Region, Environmentally and Socially Sustainable Development Department (LCSES).

Legesse, Y., Tegegn, A., and Belachew, T. (2007). Knowledge, attitude and practice about Malaria transmission and its preventive measures among households in urban areas of Assosa Zone, Western Ethiopia. *Ethiopian Journal of Health Development*, 21, 157–165.

Lozano, R., Naghavi, M., Foreman, K., Lim, S., Shibuya, K., Aboyans, V. ... and AlMazroa, M. A. (2013). Global and regional mortality from 235 causes of death for 20 age groups in 1990 and 2010: a systematic analysis for the Global Burden of Disease Study 2010. *The Lancet*, 380(9859), 2095–2128.

McSweeney, C., New, M., and Lizcano, G. (2010). The UNDP Climate Change Country Profiles: Improving the accessibility of observed and projected climate information for studies of climate change in developing countries. American Meteorological Society. DOI:10.1175/2009BAMS2826.1

Millennium Ecosystem Assessment (2005). *Ecosystems and Human Well-being: Synthesis*. Washington, DC: Island Press. Retrieved on 23 March from www.millennium assessment.org/documents/document.356.aspx.pdf

Ndaruzaniye, V. (2011). Water security in Ethiopia: Risks and vulnerabilities' assessment. Global Water Institute for Africa Climate Change, Environment and Security.

Nicholson, S. E. (2014). A detailed look at the recent drought situation in the Greater Horn of Africa. *Journal of Arid Environments*, 103, 71–79.

Norris, D. E. (2004). Mosquito-borne diseases as a consequence of land use change. *Eco-Health*, 1(1), 19–24.

Okeibunor, J. C. Orji, B. C., Brieger, W., Ishola, G., Otolorin, E. D., Rawlins, B., Ndekhedehe, E. U., Onyeneho, N., and Fink, G. (2011). Preventing malaria in pregnancy through community-directed interventions: evidence from Akwa Ibom State, Nigeria. *Malaria Journal*, 10, 227–227.

Patz, J. A., Githeko, A. K., McCarty, J. P., Hussein, S., Confalonieri, S., and De Wet, N. (2003). Climate change and infectious diseases. *Climate Change and Human Health: Risks and Responses*, 103–132.

Randell, H., Dickinson, K., Shayo, E., Mboera, L., and Kramer, R. (2010). Environmental management for malaria control: Knowledge and practices in Mvomero, Tanzania. *EcoHealth*, 7(4), 507–516.

Rogers, D. J., and Randolph, S. E. (2000). The global spread of malaria in a future, warmer world. *Science*, 289, 1763–1766.

Rylander, C., Odland, J. O., and Sandanger, T. M. (2013). Climate change and the potential effects on maternal and pregnancy outcomes: an assessment of the most vulnerable—the mother, fetus, and newborn child. *Global Health Action*, 6.

Sena, L., Wakgari, D., and Ahmed, A. (2015). Correlation of climate variability and malaria: A retrospective comparative study, Southwest Ethiopia. *Ethiopian Journal of Health Sciences*, 25(2), 129–138.

Skliris, N., Zika, J. D., Nurser, G., Josey, S. A., and Marsh, R. (2016). Global water cycle amplifying at less than the Clausius–Clapeyron rate. *Scientific Reports*, 6, 38752.

Sorenson, S. B. (2011). Safe access to safe water in low income countries: Water fetching in current times. *Social Science and Medicine*, 72(9), 1522–1526.

Stockholm International Water Institute (SIWI) (2012). Large-scale water storage in the water, energy and food nexus: Perspectives on benefits, risks and best Practices. Retrieved on 23 March 2017 from www.siwi.org/documents/Resources/Papers/Water_Storage_Paper_21.pdf

Swatuk, L. A. (2018). The land-water-food-energy nexus: green and blue water dynamics in contemporary Africa–Asia relations. In: *Routledge Handbook of Africa-Asia Relations* edited by P. M. A. Raposo de Medeiros Carvalho, D. Arase and S. Cornelissen. London: Routledge, pp. 386–405.

Swatuk, L. A. (2012). Water and security in Africa: State-centric narratives, human insecurities. In: M. A. Schnurr and L. A. Swatuk, eds, *Natural Resources and Social Conflict*. Basingstoke: Macmillan.

Swatuk, L. A., and Cash, C. (2018). *Water, Energy, Food and People Across the Global South. 'The Nexus' in an Era of Climate Change*. Palgrave: Macmillan.

Tanner, M., and Vlassoff, C. (1998). Treatment-seeking behaviour for malaria: a typology based on endemicity and gender. *Social Science and Medicine*, 46(4–5) 523–532.

Tanser, F. C., Sharp, B.le Sueur, B. D. (2003). Potential effect of climate change on malaria transmission in Africa. *The Lancet*, 362(9398), 1792–1798.

Tarhule, A. (2005). Damaging rainfall and flooding: the other Sahel hazards. *Climatic Change*, 72(3), 355–377.

Terrascope (2013). Dams and reservoirs. Retrieved on 31 March 2017 from http://12.000. scripts.mit.edu/mission2017/dams-and-reservoirs/

The World Bank (2016). Rural population (% of total population). Retrieved on 17 April 2017 from http://data.worldbank.org/indicator/SP.RUR.TOTL.ZS?locations=ET

Tilt, B., Braun, Y., and He, D. (2009). Social impacts of large dam projects: A comparison of international case studies and implications for best practice. *Journal of Environmental Management*, 90, S249-S257.

Trenberth K. E. (1998) Atmospheric moisture residence times and cycling: Implications for Rainfall rates and climate change. *Climatic Change*, 39, 667–694.

Trenberth, K. E. (2005). The impact of climate change and variability on heavy precipitation, floods, and droughts. In: M. G. Anderson, *Encyclopedia of Hydrological Sciences*. Hoboken: John Wiley and Sons.

United Nations Environment Programme (UNEP) (2006). Africa environment outlook 2: Our environment, our wealth. ISBN: 92 807 2691 9

UN Water (2013). Water and Gender. www.unwater.org/fileadmin/user_upload/ unwater_new/docs/water_and_gender.pdf

UN Africa Renewal (2014). Financing Africa's Massive Projects. www.un.org/africa renewal/magazine/december-2014/financing-africa%E2%80%99s-massive-projects

UNEP (2009). Rainwater Harvesting: A lifeline for human well-being www.unwater.org/ downloads/Rainwater_Harvesting_090310b.pdf

UNEP (2011). Food security in the Horn of Africa: The implicaitons of a drier, hotter and more crowded future. Available at: https://na.unep.net/geas/getUNEPPageWith ArticleIDScript.php?article_id=72 accessed 16 May 2018.

Velpuri, N. M., and Senay, G. B. (2012). Corrigendum to 'Assessing the potential hydrological impact of the Gibe III Dam on Lake Turkana water level using multi-source satellite data' published in *Hydrol. Earth Syst. Sci.*, 16, 3561–3578, 2012. *Hydrology and Earth System Sciences*, 16(10), 3645.

Verhoeven, H. (2013). The politics of African energy development: Ethiopia's hydroagricultural state-building strategy and clashing paradigms of water security. *Philosophical Transactions of the Royal Society of London A: Mathematical, Physical and Engineering Sciences*, 371(2002), 20120411.

Virginia Tech (2014). The grand Ethiopian renaissance dam: Sustainable development of not? Retrieved on 17 April 2017 from https://scholar.vt.edu/access/content/ group/5b95dc6f-a3ef-4ce5-8e1a-875819148663/Web/Student%20Publications/ ASwanson.GERD_final.pdf

Washington State University (2012). New global warming culprit: Methane emissions jump dramatically during dam drawdowns. *ScienceDaily*. Retrieved 17 April 2017 from www.sciencedaily.com/releases/2012/08/120808081420.htm

Watts, R. J., Richter, B. D., Opperman, J. J., and Bowmer, K. H. (2011). Dam reoperation in an era of climate change. *Marine and Freshwater Research*, 62(3), 321–327.

Williams, P., Martina, A., Cumming, R., and Hall, J. (2009). Malaria prevention in sub-Saharan Africa: A field study in Rural Uganda. *Journal of Community Health*, 34(4), 288–294.

World Bank Group (2015). Ethiopia poverty assessment 2014. Washington, DC. © World Bank. Retrieved on 17 April 2017 from https://openknowledge.worldbank.org/handle/10986/21323 License: CC BY 3.0 IGO

World Commission on Dams (2000). *Dams and Development: A New Framework for Decision-making: the Report of the World Commission on Dams*. London: Earthscan. Retrieved on 31 March 2017 from www.internationalrivers.org/sites/default/files/attached-files/world_commission_on_dams_final_report.pdf

World Health Organization (WHO) (2014). A global brief on vector-borne diseases. Geneva, Switzerland: World Health Organization. Retrieved on 31 March 2017 from http://apps.who.int/iris/bitstream/10665/111008/1/WHO_DCO_WHD_2014.1_eng.pdf

World Water Assessment Programme (WWAP) (2012). The United Nations World water development report 4: Managing water under uncertainty and risk. Paris, UNESCO. Retrieved on 31 March 2017 from http://unesdoc.unesco.org/images/0021/002171/217175e.pdf

World Wildlife Fund (WWF) (2013). Seven sins of dam building. Germany. Retrieved from http://awsassets.panda.org/downloads/wwf_seven_sins_of_dam_building.pdf

Wu, X., Lu, Y., Zhou, S., Chen, L., and Xu, B. (2016). Impact of climate change on human infectious diseases: Empirical evidence and human adaptation. *Environment International*, 86, 14–23.

Yewhalaw, D, Legesse, W., Van Bortel, W., Gebre-Selassie, S., Kloos, H., Duchateau, L., and Speybroeck, N. (2009). Malaria and water resource development: the case of Gilgel-Gibe hydroelectric dam in Ethiopia. *Malaria Journal*, 8(1), 21.

Yewhalaw, D, , Getachew, Y., Tushune, K., W/Michael, K., Kassahun, W., Duchateau, L., and Speybroeck, N.. (2013). The effect of dams and seasons on malaria incidence and anopheles abundance in Ethiopia. *BMC Infectious Diseases*, 13(1), 161.

Yewhalaw, D., Kelel, M., Getu, E., and Wessell, G. (2014). Blood meal sources and sporozoite rates of Anophelines in Gilgel-Gibe dam area, Southwestern Ethiopia. *African Journal of Vector Biology*.

Afterword
From backdraft to boomerang

Geoffrey D. Dabelko and Meaghan Parker

The road to hell is famously paved with good intentions. The battle to confront the climate challenge will be no exception – and yet, researchers and scholars have generally neglected to investigate how our efforts to mitigate and adapt to climate change might directly and immediately contribute to conflict. To avoid doing harm by doing good, researchers and policymakers must identify the risks involved in climate programs and policies. This overlooked field of research – which falls under the umbrella of 'backdraft' (Dabelko *et al.*, 2013) and now 'boomerang' (this collection, especially Chapter 1) – requires paying attention to the ways in which our responses to climate change will produce winners and losers. Otherwise, our good intentions may lead us down a dangerous path.

From threat multiplier to backdraft and boomerang

The scholarly and practitioner debate over the links connecting the environment and natural resources to conflict and security spiked following the end of the Cold War in the late 1980s. Claims that resource scarcity contributes to sub-national violent conflict shared the stage with broader efforts to redefine security by placing the environment at the heart of a hoped-for evolution away from narrow state-centred conceptions.

At the time, climate change was largely absent from these debates, due to its perceived slow onset, and diffused, undirected and uncertain impacts. But beginning in 2007, climate change became the primary entry point, as its role as a 'threat multiplier' grew to dominate military and intelligence assessments and many civilian efforts as well. Redefining security morphed into environmental links to geopolitics and state stability. The wider lens brought in previously overlooked geographies, such as the Arctic, and included regional and even global scales. The rise of climate-security reached its scientific apogee when it was included – for the first time – in the 5th Intergovernmental Panel on Climate Change assessment, represented by the 'Human Security' chapter (Adger *et al.*, 2014).

While the diversity of concerns surrounding climate–security linkages brought in a larger number of new actors and issues (particularly migration), the threat frame largely remained intact. What threats did climate changes present from the local to the global level? Climate change and conflict connections remain focused largely on how changes in water or food availability contribute to the outbreak of conflict and instability. In this way, today's research and practitioner questions differ little from the 1990s debates, even as the methods have become more diverse and the data more abundant, of better quality and of higher resolution.

Throughout this environmental security discourse, the ways the world's climate responses – both efforts to mitigate and adapt – might directly and immediately contribute to conflict and wider negative governance outcomes have been largely overlooked and underexplored. Even the most well-intentioned climate responses can neglect social, economic and equity considerations, possibly engendering grievances and destabilizing communities. These 'backdraft' dynamics have been recognized in the IPCC 5th assessment (Adger *et al.*, 2014), policy forums, such as the UN High Commission for Human Rights and the G7 Foreign Ministers (Rüttinger *et al.*, 2015) and among local environmental activists witnessing boomerang dynamics on the ground (Siakor, 2014).

Tomorrow's problems, today

Numerous well-established interventions have had demonstrable negative impacts. For example, dams and large-scale water infrastructure have a history of generating considerable negative impacts for local peoples (Conca, 2005). Changing land-tenure and natural resource rights in service of wider objectives have often come at high costs to indigenous or out-of-favour groups like pastoralists (Peluso, 1993). These familiar interventions resemble legitimate climate response options and therefore should be instructive.

Practitioners are waking up to the notion that they must 'go in with their eyes open' on climate responses. Yet backdraft and boomerang concerns are still in danger of being viewed as tomorrow's problem, not today's.

It is fair to say a backdraft argument remains largely deductive. The possible panoply of negative outcomes from trying to address climate change through mitigation or adaptation is predicated on a wide range of actors actually making significant mitigation and adaptation efforts. Despite the fanfare and cautious optimism from the 2015 Paris agreement, the international community is much farther along in diagnosing climate change problems than in taking fundamental steps to address them. Even with complete fulfilment of the Paris agreement's commitments, the global metrics of climate progress – parts per million (PPM) of carbon dioxide in the atmosphere and total average temperature changes – are likely to fall short of avoiding catastrophic outcomes.

Hence, those pursuing backdraft and boomerang questions face pushback when they seek to prioritize these concerns. To wave red flags against climate change action prematurely may undermine the efforts to produce meaningful climate responses in the first place. However, this argument is flawed in at least two ways.

First, the goal of identifying backdraft dynamics is to inform climate change responses so that these unintended effects can be minimized and the trade-offs producing relative winners and losers can be recognized up front before the harm is done. A deductive approach – before the dynamics are fully manifested – thus has proactive and preventive utility, rather than an entirely reactive character. Avoiding misguided or costly approaches not only reduces the negative impacts, it also helps minimize the chance that these mistakes could delegitimize critical tools and processes for climate response.

Payment for ecosystem services, for example, is a critically important approach, but pursuing REDD+ schemes that trample human rights, line the pockets of elites and lead to legitimate grievances against climate mitigation schemes damages more than the immediate communities. It could also lead to loss of confidence in the mechanisms themselves; they could be seen as ineffective, poorly executed and too expensive. This dynamic is demonstrated by the REDD readiness programme in Tanzania (Beymer-Farris and Bassett, 2012), where the corrupt local implementer and the lack of sensitivity to the pre-existing and unrelated conflict between the national government and local peoples combined to create a dysfunctional and conflictual project. Such examples present lessons that – if examined and shared – can help other payment for ecosystem services schemes operate more fairly and with less structural violence.

Second, a truly effective response to climate change is not required to generate highly undesirable impacts. The negative poverty, access and equity dynamics outlined by Larry Swatuk, Lars Wirkus and their colleagues in this collection (Chapter 1) make clear that the boomerang effect can play out even with relatively ad hoc, modest and pilot climate change mitigation and adaptation efforts.

Local, national and international level efforts to grapple with climate change are unsteady, underdeveloped and insufficient. Some actors continue to deploy tools of obfuscation, delay and opposition to climate change response. But even with these insufficient efforts, considerable climate action is underway in multiple sectors. Thus the time is now to implement the precautionary principle – not only for climate impacts, but for climate action as well.

Busting the climate silo

The biggest barrier to preventing conflict from climate responses is self-imposed. The impacts of climate change, and these responses to it, are still too narrowly framed, remaining the purview of the 'climate community' or a single appointed agency, department, NGO or business identified as responsible for climate change (which is never the entity responsible for peace and security). There have been some promising integrated and transdisciplinary analysis and action in certain arenas, but it is still extremely limited.

The persistent and dangerous stove-piping will continue to exacerbate backdraft and produce boomerang effects. The knock-on effects of failing to account for poverty, justice and access dimensions of climate responses would be far more visible if integrated and transdisciplinary approaches become the norm rather than the exception. There are some concrete ways to begin that build off of existing initiatives:

- require conflict-sensitivity screens for all climate programs throughout the project lifecycle as some bilateral aid agencies are doing;
- create more flexible pools of funding that support multiple inflection points to meeting the climate challenge that emphasize local and affected community participation;
- expand the universe of 'appropriate' climate responses and 'appropriate' climate actors to bring the social, political, economic and equity concerns into decision-making from the outset; and
- establish more applied research and practitioner forums for shared learned on cross-sectoral approaches to complex climate-security challenges.

Busting these silos is no easy task, but these actions are modest first steps to making an essential transition to how we frame and address climate change challenges.

References

Adger, W. N., J. M. Pulhin, J. Barnett, G. D. Dabelko, G. K. Hovelsrud, M. Levy, Ú. Oswald Spring and C. H. Vogel (2014). Human security. In: C. B. Field, V. R. Barros, D. J. Dokken, K. J. Mach, M. D. Mastrandrea, T. E. Bilir, M. Chatterjee, K. L. Ebi, Y. O. Estrada, R. C. Genova, B. Girma, E. S. Kissel, A. N. Levy, S. Mac-Cracken, P. R. Mastrandrea and L. L. White, eds, *Climate Change 2014: Impacts, Adaptation, and Vulnerability. Part A: Global and Sectoral Aspects. Contribution of Working Group II to the Fifth Assessment Report of the Intergovernmental Panel on Climate Change*. Cambridge: Cambridge University Press, pp. 755–791.

Beymer-Ferris, B. A. and T. J. Bassett (2012). The REDD menace: resurgent protectionism in Tanzania's mangrove forests. *Global Environmental Change*, 22(2), 332–341.

Conca, K. (2005). *Governing Waters: Contentious Transnational Politics and Global Institution Building*. Cambridge, MA: MIT Press.

Dabelko, G. D., L. Herzer, S. Null, M. Parker and R. Sticklor (eds) (2013). Backdraft: the conflict potential of climate change adaptation and mitigation. *Environmental Change and Security Program Report* 14(2).

Peluso, N. L. (1993). Coercing conservation? The politics of state resource control. *Global Environmental Change*, 3(2), 199–217.

Rüttinger, L., D. Smith, G. Stang, D. Tänzler, J. Vivekananda, O. Brown, A. Carius, G. D. Dabelko, R. D. Souza, S. Mitra, K. Nett, M. Parker and B. Pohl (2015). *A New Climate for Peace: Taking Action on Climate and Fragility Risks*. Berlin: adelphi, International Alert, Wilson Center, European Institute for Security Studies.

Siakor, S. K. A. (2014). The real price of Europe going green. In: K. Conca and G. D. Dabelko, eds, *Green Planet Blues: Critical Perspectives on Global Environmental Politics* (5th edn). Boulder, CO: Westview Press, pp. 359–363.

Index